Index Theorem. 1

IWANAMI SERIES IN MODERN MATHEMATICS

Translations of
MATHEMATICAL
MONOGRAPHS

Volume 235

Index Theorem. 1

Mikio Furuta

Translated by
Kauro Ono

American Mathematical Society
Providence, Rhode Island

指数定理 1

(INDEX THEOREM 1)

by Mikio Furuta

Originally published in Japanese
by Iwanami Shoten, Publishers, Tokyo, 1999
Translated from the Japanese by Kaoru Ono

2000 *Mathematics Subject Classification.* Primary 58–01, 58J20.

For additional information and updates on this book, visit
www.ams.org/bookpages/mmono-235

Library of Congress Cataloging-in-Publication Data

Furuta, M.
Index theorem. 1 / Mikio Furuta ; translated by Kaoru Ono.
p. cm. — (Translations of mathematical monographs ; v. 235) (Iwanami series in modern mathematics)
Includes bibliographical references and index.
ISBN 978-0-8218-2097-1 (alk. paper)
1. Index theorems.

QA614.92.F87 2007
514′.74—dc22 2007060301

10 9 8 7 6 5 4 3 2 1   12 11 10 09 08 07

Contents

Preface

Linear differential equations are looked upon from two points of view.

The first point of view is based on understanding a linear differential equation as a linear mapping between two infinite-dimensional vector spaces. Let us consider the situation when the linear mapping gives an "almost" linear isomorphism. Here, "almost" means "ignoring finite-dimensional vector subspaces." In other words, the finite-dimensional subspaces, which are ignored, are responsible for the possibility that the two infinite-dimensional vector spaces are not canonically isomorphic. This situation yields a linear mapping between certain finite-dimensional vector subspaces. The object obtained by using this finite-dimensional approximation is called the index.

Another point of view is based on the description of a linear differential equation using local coordinates. Then the linear differential equation can be formally regarded as a section of a certain fiber bundle. The object obtained using this approach is the homotopy class of the section and is called (the homotopy class of) the principal symbol.

Each object is invariant under continuous deformation, with appropriate requirements, of the linear differential equation. In this sense, both of them are topological objects.

The two points of view are related to each other. The condition that the principal symbol is almost invertible is a sufficient condition for the linear differential operator to be an almost linear isomorphism (at least on closed manifolds). The former is called ellipticity for partial differential equations, and the latter is called Fredholm property, which is a notion in functional analysis. Ellipticity implies Fredholm property.

In order to compute the index following its definition, we must solve the linear differential equation to obtain a finite-dimensional approximation. Among various approaches to the index, the most basic one is given by the integer

$$\operatorname{ind} P = \dim \operatorname{Ker} P - \dim \operatorname{Coker} P,$$

where we denote the linear mapping by P.

The Atiyah-Singer index theorem describes how to determine the index of an elliptic differential operator by its principal symbol using only topological tools.

As a slogan, the index theorem states the coincidence of the analytical index, which is the original definition, and the topological index, which is defined using topological tools.

The index theorem as a coincidence of two numbers originated in the work of Atiyah and Singer. Its applications to differential topology have also been explored since then, but this area of research is not active at this moment. In recent years, the index theorem has become rather a daily tool used in global analysis involving non-linear differential equations. A few decades ago, the index theorem itself was the focal point of the formation of global analysis.

Is research on the index theorem "done with"?

The author believes that this is not the case. One piece of evidence is that there are still active research fields surrounding the index theorem. Outside of such obviously related places, it seems to the author that the study of each subject in the background of the index theorem is still in progress. It is not yet clear what future theory the index theorem uses as its first step, but surely it will be something interesting. The contours of this future theory can be vaguely seen when we review the development of the index theorem.

At this moment, it is important to understand the index theorem not as a mountain chain, behind grand theories, but to digest it as a natural element of the scenery when we follow the fundamental properties of differential equations.

I would like to present such understanding of the index theorem by giving proofs that are as elementary as possible. This is what I always had in mind when writing this book. I would be happy if readers would give opinions and criticism.

A part of this book was written during the author's stay at the Max-Planck-Institut für Mathematik. I appreciate the institute for providing such a good working environment.

October, 1998

Mikio Furuta

Outline of the Theory and Perspective

de Rham complex and Dolbeault complex

The de Rham complex and the Dolbeault complex are two complexes that appear in the geometry of differentiable manifolds and the geometry of complex manifolds, respectively.

The former, resp. the latter, has its origin in Cartan's discovery of the exterior derivative, resp. Cauchy's discovery of the Cauchy-Riemann equation.[1] In the former case, its connection with geometry was revealed when the de Rham theorem clarified its relation to Betti numbers. In the latter case, it goes back to an older time, when Riemann found the ancestor of the Riemann-Roch theorem, i.e., the inequality involving the genus of the Riemann surface. However, the former was understood as the device deriving invariants of manifolds and the latter was understood as some properties of various holomorphic line bundles (or divisors). Hence they play different roles. This is still the case now.

These two trends met in a clear form under the viewpoint of unification in the theory of Hodge and Kodaira, which followed Weyl's work. Cohomology is associated with a complex and has geometric meaning. (For clear understanding, it was also necessary to establish the notion of sheaves and their Čech cohomology, which is equivalent to the two kinds of complexes mentioned above.) Hodge-Kodaira described their cohomologies in terms of solutions of the equation of Laplace type.

Considering the alternating sum of the dimension of the cohomology of the complex instead of the dimension of the cohomology in each degree, we lose a lot of detailed information, but we acquire simple properties (with respect to exact sequences). This fact is clear

[1] Manifesting the origin may depend on personal opinion. Other descriptions in this part are also like virtual history.

in an algebraic setting. However, the alternating sums for the two complexes have developed in different ways.

For the de Rham complex, its alternating sum is called the Euler characteristic. As its name indicates, the Euler characteristic itself was known a long time ago. However, it was after Poincaré started the subject, which is now called topology, and then considered homology, that the geometric meaning of the Euler characteristic began to be clarified. The Poincaré-Hopf theorem is famous in its relation to the number of zeros of a vector field on a manifold. This theorem states that the Euler characteristic can be expressed as the integration of a characteristic class called the Euler class (in general, the integration of characteristic classes are called characteristic numbers).

For the Dolbeault complex (or the corresponding Čech complex), it was already a historical achievement to construct the higher cohomology groups as the objects for taking the alternating sum. The Riemann-Roch theorem gives the formula for the alternating sum in the case of complex dimension 1. Based on results for low-dimensional cases, such as complex surfaces, due to Kodaira and others, it was Hirzebruch who finally gave the formula for the general case (nonsingular projective algebraic varieties). The end result is given as the combination of various characteristic numbers.

Vaguely speaking, the alternating sum is invariant under quite a large class of deformations of the complex. In order to investigate properties of the alternating sum by use of the freedom of such deformations, it is necessary to make the precise meaning of "deformation" that would be as broad as possible.

From a modern viewpoint, the ideas of "derived categories" (in algebraic geometry) and "stable homotopy" (in topology) provide a framework in which equivalence classes under deformations (with certain freedom of deformations) are described. It was Grothendieck who introduced cohomology (of sheaves) in algebraic geometry suitable for the former notion. In particular, in our current problem on the alternating sum of the dimensions of cohomology groups of the complex, the notion of the K-group, which naturally emerges from these frameworks, is useful. When Grothendieck extended the Riemann-Roch theorem to families of algebraic varieties of arbitrary characteristic, rather than to a single algebraic variety, he used the notion of the K-group. Following such developments, the topological version of the K-group was later introduced by Atiyah as the set of a certain kind of stable homotopy classes.

The freedom that Hirzebruch could use for extending the Riemann-Roch theorem was rather limited as compared to the freedom Grothendieck later enjoyed when he invented a very general approach to algebraic geometry and used it for the above-mentioned extension of the Riemann-Roch theorem. Instead, Hirzebruch used, in an essential step, a deep theorem, which belongs to differential topology. It is the signature theorem, which is also due to Hirzebruch himself.

Hirzebruch's signature theorem

Hirzebruch's signature theorem is purely a theorem in differential topology. Similar to the fact that the Euler characteristic partially reflects the information on the dimension of the de Rham cohomology of the manifold, the signature partially reflects the information on the cup product on the de Rham cohomology. This theorem describes the signature of the manifold explicitly using characteristic numbers.

In order to prove the signature theorem, which is formulated for each (oriented) closed manifold, Hirzebruch did not discuss each individual manifold, but used Thom's cobordism theory, which deals with all closed manifolds at once.

Consider the set of all diffeomorphism classes of closed manifolds, which are not necessarily connected, for all dimensions. To introduce the algebraic structure on this set, define the sum by the disjoint union as spaces and the product by the product as spaces. Take the quotient by the relation defined by the requirement that a manifold, which is the boundary of some compact manifold, is equivalent to 0. This is the cobordism ring. In fact, this ring is a commutative ring (in the sense of graded rings) with the identity being the empty set. We can also define the cobordism ring of closed manifolds, or closed manifolds equipped with any other suitable structure. On the other hand, the totality of characteristic classes has a structure of a commutative ring. The characteristic numbers are obtained by the integration of the characteristic classes of the tangent bundle over the manifold. This integration defining the characteristic numbers yields a bilinear form between the cobordism ring and the ring of characteristic classes. Thom's theory claims that this bilinear form is non-degenerate in the case of the cobordism ring of oriented closed manifolds after tensoring $\mathbb{Q}$.

Hirzebruch noticed that the signature gives a ring homomorphism from the cobordism ring to $\mathbb{Z}$. By Thom's theory, this immediately implies that the signature is expressed in terms of characteristic numbers. Once the explicit expression is predicted, it is easily verified by confirming the prediction for generators of the cobordism ring.

How does the proof of Thom's cobordism theory go? Logically, it consists of the following four steps.

(1) Determine the ring of all characteristic classes (the point is that any vector bundle is realized as a subbundle of a trivial vector bundle. The stable homotopy class of the embedding is uniquely determined. It reduces the problem to the computation of the cohomology of the Grassmann variety (the classifying space)).
(2) Estimate the size of the cobordism ring from below (list all candidates for generators. In practice, candidates are given by projective spaces of all dimensions).
(3) Estimate the size of the cobordism ring from above (the point is that any manifold is realized as a closed subspace of the Euclidean space. The Pontrjagin-Thom construction reduces the problem to the computation of the homotopy of the space called the Thom complex).
(4) Observe that the bilinear form between (1) and (2) is non-degenerate and that the size in (3) is at most the size of (1).

Then we find that (2) and (3) are equal and that the bilinear form between (1) and (3) must be non-degenerate. This is the desired statement.

The first two proofs of the index theorem

Using the de Rham cohomology and the Hodge theory, the signature is expressed as the dimension of Ker minus that of Coker, namely the "index" of a certain elliptic differential operator.

Similarly, the Poincaré-Hopf theorem related to the de Rham complex and the Riemann-Roch theorem related to the Dolbeault complex are interpreted as the formulas of the index of elliptic linear differential opeators, which are obtained by folding the $\mathbb{Z}$-graded complexes to $\mathbb{Z}_2$ ones.

Atiyah-Singer generally proved the formula for the index of elliptic linear differential operators in terms of the combination of characteristic numbers, which includes the above three cases. This is the "index theorem".

The first proof of the index theorem was the extension of Hirzebruch's argument using Thom's cobordism theory.

Then Atiyah-Singer announced the second proof of the index theorem. In this proof, a more direct framework was given using the (topological) K group beforehand of the formula in terms of characteristic numbers.

The second proof contains three points.

Compared with the approaches taken by Grothendieck and Hirzebruch, where they remained in the framework of algebraic geometry and topology as much as possible, Atiyah-Singer's second proof of the index theorem uses the freedom of "pseudo-differential operators", which are closely related to the elliptic differential operators, which are the original objects of the theorem. Then the first point of the proof is to check fundamental properties of elliptic pseudo-differential operators.

Thom's cobordism theory, which Hirzebruch used, requires a meta-global argument, which captures all manifolds and all characteristic classes. From the viewpoint of studying individual manifolds, it may be considered as an indirect tool. To capture all manifolds at once, the argument is based on the fact that any manifold can be embedded in the Euclidean space of sufficiently large dimension. Atiyah-Singer's second proof of the index theorem directly used this fact as a geometric idea. It is only necessary to embed each individual manifold, and one needs not to capture all manifolds at once. This is the second point.

The third point is "Bott's periodicity theorem". It is, in the final stage, used for verifying the index theorem in the case of the Euclidean space, to which the index theorem in the general case is reduced. Atiyah pointed out that Bott's periodicity theorem is easily explained using the notion of the topological K-group and the notion of the "index for families of elliptic linear differential operators". Each of these notions were modelled on the extension of the Riemann-Roch theorem due to Grothendieck. However, the only known proof, which covers the case with (non-commutative) compact group action, is Atiyah's argument, which uses the index.

Other proofs of the Atiyah-Singer index theorem

The second proof and the formulation using the K-group therein are suitable for extending the index theorem to wider situations. For example, the formulation and the proof for the case of the index of families immediately follow.

On the other hand, there are some other approaches for the original index theorem dealing with the numerical index (or characteristic numbers).

- The method of the heat kernel.[2] The local properties, which the index enjoys, are realized by the study of the asymptotic expansion of the trace of the heat kernel. It is also a starting point toward the extension of the index theorem for manifolds with boundary due to Atiyah-Patodi-Singer.
- The proof using stochastic analysis. It is related to the method of the heat kernel. However, in this approach, we, in a sense, consider the stage before (pseudo-) differential operators, and directly study the situation when the heat kernel emerges.
- Explanation using symplectic geometry on the loop space. This consideration uses the Duistermaat-Heckman formula, which is similar to the residue formula for integration. It is an intuitive argument by Witten, but an appropriate mathematical framework has not been given yet.

Both the first and second proofs due to Atiyah-Singer fully used the topological freedom. In other regards, roughness of the mesh in such arguments is just enough for the proof and the minimal understanding of the index theorem. On the other hand, other proofs rely on more detailed geometry. We may say that these arguments do not reduce the problem to simpler cases by using the topological freedom, but may help to understand the index theorem as geometry that arises in discussion of differential operators.

It is fundamental in the index theorem, or the notion of the index, that certain finite-dimensional information is defined by cancellation between infinite-dimensional parts. We may say that each of these other proofs helps in the direct understanding of such a mechanism of cancellation.

Nowadays, when discussing the index theorem, we cannot ignore its relation to physics. The above argument due to Witten is one such example. In this book, we do not describe their relation, but only list some names of related physicists and mathematicians.

[2]Cf. Tomoyoshi Yoshida, "Index Theorem for Dirac Operators" (Kyoritsu Shuppan Publisher, 1998, in Japanese).

Alvarez-Gaumé, Witten, Getzler, Quillen, Mathai,
Bismut, Berline, Vergne, etc.

We make only one remark that there is a phenomenon, called "supersymmetry" in physics, which is a mechanism of cancellation. The place where the index theorem is related to physics is in the relation to this mechanism.[3]

Characteristic features of this book

In this book, we will present the second proof of the Atiyah-Singer index theorem as far as we can, avoiding the use of pseudo-differential operators. We also intend to explain some applications of the index theorem.

So far, we have explained the index theorem according to its time evolution. An outline of the theory will be given in Chapter 1, titled "Prelude". Chapter 1 is a "sketch", and some parts use terminology without giving the definition. Thus it is not necessary that the reader read it carefully at first. The real arguments start in Chapter 2. By looking at the corresponding part of Chapter 1 again before reading the real arguments, the outline of the theory may play the role of side remarks or comments.

In this book, we deal only with "operators of Dirac type associated with $\mathbb{Z}_2$-graded Clifford modules" rather than with all elliptic differential operators. The key is the behaviour of operators of Laplace type, which is the square of operators of Dirac type. We present the formulation of the index theorem before introducing the K-group and the characteristic classes.

Our guiding principle is the "localization". This is the phenomenon showing that the index can be captured by looking at parts of the manifold without dealing with the whole manifold. The essential part of this localization arises, prior to the properties of the index, as the naive phenomenon that eigenfunctions for the operator of Laplace type with a potential decay exponentially, whereas the potential function takes large values.

The localization to open submanifolds is formulated as the "excision theorem". The localization to closed submanifolds is studied through the tensor product of the operator of Dirac type on the closed submanifold and a certain operator along its normal direction.

[3]It is not necessarily the association with linear differential equations where the supersymmetry arises.

For the proof of the index theorem, we only use differential operators (we do not use pseudo-differential operators). The index will be defined for operators of Dirac type not only on closed submanifolds, but also on open submanifolds with certain conditions at infinity for the operator. We do not assume orientability of manifolds and do not deal with odd-dimensional manifolds. Recently, such objects have appeared in some applications (for example, the index theorem on the 3-dimensional Euclidean space is necessary for analysis of monopoles).

Operators on open manifolds play an important role not only in such applications, but also in theoretical considerations. The most fundamental role in this book is the "unit", namely the operator of Dirac type on the Euclidean space, which is called the "supersymmetric harmonic oscillator" in physics.

Prerequisites.

In geometry, we assume a basic knowledge of manifolds, differential forms and vector bundles. There are also some places where the reader will need a little knowledge in topology (e.g., covering spaces, homotopy groups, the Poincaré duality, etc.).

In algebra, we assume knowledge of linear algebra, including the exterior product, the tensor product, etc. We also use very basic facts on the representation theory of finite groups and compact groups in § 2.3 and § 4.4.

In analysis, we assume knowledge of local properties of partial differential equations on the Euclidean space. However, the place where some prerequisite in analysis in the first volume is required is only in Chapter 4 (the Rellich theorem, *a-priori* estimates, regularity for solutions of elliptic differential operators, etc.). If the reader is interested only in geometric arguments, then (s)he can skip Chapter 4 and proceed to later chapters. The second volume [4] contains the study of non-linear differential equations, and we quote some facts from functional analysis.

[4]Editor's Note: *Index Theorem. 2* is forthcoming.

Terminology and notation.

- $\mathbb{C}$, $\mathbb{R}$, $\mathbb{Z}$: the sets of complex numbers, real numbers, and integers, respectively. We also use some standard notation without mentioning them here.
- $\mathbb{Z}_2$: we abbreviate $\mathbb{Z}/2\mathbb{Z}$ in this way (it does not stand for the 2-adic numbers).
- $\operatorname{Ker} f, \operatorname{Coker} f, \operatorname{Im} f$: for a linear mapping f, we denote its kernel, cokernel and image by these symbols.
- id_E: the identity on E.
- Unless otherwise stated, "vector spaces" are finite dimensional.
- Unless otherwise stated, "manifolds" are differentiable manifolds, "vector bundles" are smooth vector bundles, and "functions" and "sections" are smooth.
- For a smooth vector bundle F on a manifold X, we denote by $(F)_x$ the fiber at $x \in X$. We also denote by $\Gamma(F)$ the set of all smooth sections of F.
- ϵ_E: the endomorphism of a $\mathbb{Z}_2$-graded vector space (or vector bundle) $E = E^0 \oplus E^1$, which coincides with id_{E^0} on E^0 and with $-\mathrm{id}_{E^1}$ on E^1.
- Since a "$\mathbb{Z}_2$-graded Clifford module with compact support" is too long, we simply call it a "pair".
- $H^k(X, \theta)$: the de Rham cohomology with coefficients in the local system θ on X.
- We tried to use notation systematically throughout the book; e.g., most of the time E stands for a vector space or Euclidean space T a vector bundle with Euclidean metric, F a vector bundle, which we take the tensor product as coefficients, but we could not avoid coincidences of notation. For example, F_A is the symbol for the curvature and may be confused with the symbol F for a vector bundle. The reader is asked to be careful with such cases.

CHAPTER 1

Prelude

The index is the information extracted from the linear mapping between infinite dimensional vector spaces, which comes from linear differential operators, so that it is invariant under "small perturbations". The index theorem provides the way of computing the index. It is easy to formulate and prove the index theorem in the one-dimensional case, so we start by explaining it.

1. What Is the Index?

1.1. Definition of the Index. Let $f : E^0 \to E^1$ be a linear mapping between finite dimensional vector spaces E^0 and E^1. Then we have

$$\dim \operatorname{Ker} f - \dim \operatorname{Coker} f = \dim E^0 - \dim E^1.$$

We interpret this fact in two ways.

(1) Suppose that we want to compute the difference of the dimension of the vector spaces E^0 and E^1. If a linear mapping f is given, it helps us with the computation as follows. The key point is that f induces an isomorphism from any complementary subspace F of $\operatorname{Ker} f$ to the image $\operatorname{Im} f$. In particular, their dimensions are the same so that they cancel each other when we consider the difference between the dimensions. Hence it is enough to compute the difference between the dimensions of $\operatorname{Ker} f$ and the complementary subspace of $\operatorname{Im} f$.

(2) The dimension of $\operatorname{Ker} f$ and $\operatorname{Coker} f$ depend on the mapping f. But the difference between the dimensions is independent of f. Therefore it gives a quantity, which carries a basic feature, like a "spine" of the mapping f.

These two points are obvious when we deal with finite dimensional vector spaces. The feature, however, becomes different when we deal with infinite dimensional vector spaces.

Let $P : \mathcal{E}^0 \to \mathcal{E}^1$ be a linear mapping between infinite dimensional vector spaces $\mathcal{E}^0$ and $\mathcal{E}^1$. Then the argument (1) above does not make sense, since each term of

$$\dim \mathcal{E}^0 - \dim \mathcal{E}^1$$

is infinity and does not make sense as it is.

How is the argument (2) above? Suppose that Ker P and Coker P are finite dimensional, so that the difference

$$\dim \operatorname{Ker} P - \dim \operatorname{Coker} P$$

is defined as an integer. In such a case, P is called **Fredholm**. However, the difference above depends on P.

For example, let us consider the set of all sequences of real numbers indexed by integers $(a_k)_{k\in\mathbb{Z}}$. This set is a real vector space in an obvious way. We define linear mappings P_0 and P_1 from $\mathcal{E}^0$ to $\mathcal{E}^1$, where $\mathcal{E}^0 = \mathcal{E}^1 = \{(a_k)_{k\in\mathbb{Z}} | a_k \in \mathbb{R}\}$.

Let P_0 be the identity mapping. Then Ker $P_0 = \{0\}$ and Coker $P_0 = \{0\}$, thus we have

$$\dim \operatorname{Ker} P_0 - \dim \operatorname{Coker} P_0 = 0.$$

On the other hand, we define P_1 as follows. Let (a_k) be a sequence as above. For $k < 0$, we put $b_k = a_k$. For $k \geq 0$, we put $b_k = a_{k+1}$. Then we define

$$P_1((a_k)_{k\in\mathbb{Z}}) = (b_k)_{k\in\mathbb{Z}}.$$

Then we have

$$\operatorname{Ker} P_1 = \mathbb{R}(\cdots, 0, 0, 1, 0, 0, \cdots), \quad \operatorname{Coker} P_1 = \{0\},$$

and, hence,

$$\dim \operatorname{Ker} P_1 - \dim \operatorname{Coker} P_1 = 1.$$

Comparing P_0 and P_1, the difference between them is caused by "infinite dimensionality" in the sense that the infinitely many basis elements are shifted. The difference between the dimension of the kernel and the dimension of the cokernel is invariant when we perturb P_0 in "finite dimensional way". More precisely, it is an exercise in linear algebra to show that, for a Fredholm linear mapping $P : \mathcal{E}^0 \to$

$\mathcal{E}^1$, $P+Q$ is also a Fredholm linear mapping, where $Q : \mathcal{E}^0 \to \mathcal{E}^1$ is a linear mapping with the finite dimensional image, and satisfies

$$\dim \operatorname{Ker}(P+Q) - \dim \operatorname{Coker}(P+Q) = \dim \operatorname{Ker} P - \dim \operatorname{Coker} P.$$

From consideration of the above, neither of two interpretations can work directly in the infinite dimensional case. However, this discussion provides a strategy.

(1) Suppose that we would like to define something like differences between dimensions of infinite dimensional vector spaces $\mathcal{E}^0$ and $\mathcal{E}^1$. When a Fredholm mapping P is given, we can use it and proceed as follows. The key point is that the restriction of P to any complementary subspace $\mathcal{F}$ to $\operatorname{Ker} P$ is an isomorphism onto $\operatorname{Im} P$. We regard that they cancel each other. Then the difference between the dimensions of the remaining parts (complementary subspaces) can be taken as the definition of the difference between the dimensions of the total spaces.
(2) Dimensions of $\operatorname{Ker} P$ and $\operatorname{Coker} P$ depend on a Fredholm mapping P. But the difference between these dimensions hardly changes under "small perturbation" of P. For example, it is invariant under addition of Q, where Q is a linear mapping with finite dimensional image. Hence this quantity must be considered as a basic property of P, something like a "spine".

In this way, we can define something like the difference between the dimensions of $\mathcal{E}^0$ and $\mathcal{E}^1$ for a triple $(\mathcal{E}^0, \mathcal{E}^1, [P])$. Here, P is a Fredholm operator and $[P]$ is an equivalence class, which only captures its "spine".

Conversely, the supplementary datum which is necessary for defining the difference between the dimensions of infinite dimensional vector spaces is provided by a certain equivalence class of P.

What is the definition of an appropriate equivalence relation? We, at least, want P and $P+Q$ to belong to the same equivalence class. If we require only this property, the equivalence class is too narrow.

There is a framework for dealing with such a problem in the most general way, which is provided by functional analysis.

For example, suppose that the norms are equipped to $\mathcal{E}^0$ and $\mathcal{E}^1$ so that they are complete, i.e., they are Banach spaces. We assume that P is continuous (in other words, bounded). Then the closure of the set of linear mappings with finite dimensional image is the set of compact operators, namely, operators which map a bounded set to a relatively compact set. In another respect, the operator norm measures the difference between operators, and we have one way of making precise the meaning of "small perturbation". In this situation, we can prove that the difference between the dimension of the kernel and the cokernel is invariant under continuous deformation through Fredholm operators. Then, in this functional analytical framework, we get the best possible definition for the equivalence relation given by the condition that operators belong to the same connected component of the space of Fredholm operators. Also, Fredholm operators are simply characterized as those which induce isomorphisms between Banach spaces up to compact operators[1].

REMARK 1.1. If the difference between the dimension of the kernel and the dimension of the cokernel is not zero, this implies existence of non-zero elements either in the kernel or the cokernel. In analysis, to show existence, without explicit construction, in an abstract way, some completeness condition is usually required. (Recall the relation between the mean value theorem and the completeness of real numbers.) By requiring the completeness for norm spaces, the property of Fredholmness is preserved under continuous deformation.

As a theory, it is simple and beautiful. It is also useful, in practice, to apply this functional analytical framework, when we deal with concrete examples of Fredholm operators appearing in analysis. Moreover, there is a subject[2] which discusses how much of geometry can be reconstructed by a thorough procedure in such a framework, beyond just a generalization, and tries to make a new sense of geometry.

It is, however, possible to investigate the operators more concretely than general theory in functional analysis, when they arise as differential operators on manifolds. The Atiyah-Singer index theorem, which we study in this book, concerns such concrete Fredholm operators. Explicit consideration on concrete Fredholm operators enables us to compute the difference between dimensions of infinite dimensional vector spaces.

[1] There exists P' such that $P'P - I$ and $PP' - I$ are compact operators.

[2] Non-commutative differential geometry is such a subject.

Before presenting a bit more explanation in this section, we conclude the general framework by returning to the arguments (1) and (2).

Summing up the arguments (1) and (2) in a different way:

For a Fredholm linear mapping $P : \mathcal{E}^0 \to \mathcal{E}^1$, we approximate it by finite dimensional object as follows.

Namely, we decompose them into direct sums:

$$\mathcal{E}^0 = E^0 \oplus \mathcal{F}^0, \quad \mathcal{E}^1 = E^1 \oplus \mathcal{F}^1, \quad P = f \oplus P',$$

where E^0 and E^1 are finite dimensional, and P' is an isomorphism between infinite dimensional vector spaces. (We can take any such decomposition in the consideration.) Then this f is a "finite dimensional approximation" of P, and

$$\dim E^0 - \dim E^1$$

is independent of the choice of such a decomposition and determined only by P.

If we take $E^0 = \operatorname{Ker} P$ and $E^1 = \operatorname{Coker} P$, this reduces to the argument (1) before. Also, a proof of (2) is given by choosing common E^0 and E^1 for both of P and $P + Q$.

Now, we define the index.

DEFINITION 1.2. For a Fredholm operator P, we define its **index** $\operatorname{ind} P$ by

$$\operatorname{ind} P = \dim E^0 - \dim E^1,$$

where E^0 and E^1 are finite subspaces appearing in a finite dimensional approximation $f : E^0 \to E^1$. It does not depend on the choice of finite dimensional approximations. In particular, we have

$$\operatorname{ind} P = \dim \operatorname{Ker} P - \dim \operatorname{Coker} P.$$

In the framework of Fredholm operators between Banach spaces, this definition must be fundamental. On the other hand, when we work with Hilbert spaces (hence, equipped with an inner product), we may deal $\operatorname{Ker} P$ and $\operatorname{Coker} P$ in an equal way as follows.

Firstly, we consider finite dimensional situation $f : E^0 \to E^1$, where E^0 and E^1 are equipped with inner products. Then we have the adjoint mapping $f^* : E^1 \to E^0$ and $\operatorname{Coker} f$ is isomorphic to $\operatorname{Ker} f^*$. We also have

$$\operatorname{Ker} f = \operatorname{Ker} f^* f, \quad \operatorname{Ker} f^* = \operatorname{Ker} f f^*,$$

hence,

$$\begin{aligned}\dim \operatorname{Ker} f - \dim \operatorname{Coker} f* &= \dim \operatorname{Ker} f - \dim \operatorname{Ker} f^* \\ &= \dim \operatorname{Ker} f^* f - \dim \operatorname{Ker} f f^*.\end{aligned}$$

Let us put it in a formal way.

Take a direct sum decomposition $E = E^0 \oplus E^1$. In order to keep the decomposition in mind, we introduce a self-adjoint operator ϵ_E, which is the identity on E^0 and -1 times the identity on E^1. Then E^0 and E^1 are recovered as eigenspaces of ϵ_E. We call such a structure a $\mathbb{Z}_2$-graded vector space.

Now we put

$$\tilde{f} = \begin{pmatrix} 0 & f^* \\ f & 0 \end{pmatrix},$$

then it is a symmetric transformation anti-commuting with ϵ_E. (If we work with complex vector spaces with Hermitian inner products, we have a Hermite transformation.) Namely,

$$\tilde{f}\epsilon_E = -\epsilon_E \tilde{f}.$$

We say that an endomorphism operator, which is anti-commutative with ϵ_E, has **degree 1**.

The operator

$$\tilde{f}^2 = \begin{pmatrix} f^* f & 0 \\ 0 & f f^* \end{pmatrix}$$

is a symmetric transformation commuting with ϵ_E (when we work with complex vector spaces, it is a Hermitian transformation) whose eigenvalues are all non-negative. We say that an endomorphism operator, which commutes with ϵ_E, has **degree 0**.

In this way, after introducing $\tilde{f}$, we have

$$\dim \operatorname{Ker} f - \dim \operatorname{Coker} f^* = \dim \operatorname{Ker}(\tilde{f}^2|E^0) - \dim \operatorname{Ker}(\tilde{f}^2|E^1).$$

Decompose the eigenvalues according to the decomposition $E = E^0 \oplus E^1$, the multiplicities are the same except for the zero eigenvalue. In fact, f and f^* give isomorphisms between these eigenspaces. Based on this fact, we also get

$$\dim \operatorname{Ker}(\tilde{f}^2|E^0) - \dim \operatorname{Ker}(\tilde{f}^2|E^1) = \dim E^0 - \dim E^1.$$

From the consideration above, we formulate the notion of the index as follows.

DEFINITION 1.3. Let $\mathcal{E} = \mathcal{E}^0 \oplus \mathcal{E}^1$ be a $\mathbb{Z}_2$-graded (infinite dimensional) real vector space with an inner product and $\tilde{P}$ a symmetric operator of degree 1 such that $\operatorname{Ker} \tilde{P}^2$ is finite dimensional. Then we define the **index** $\operatorname{ind} \tilde{P}$ by

$$\operatorname{ind} \tilde{P} = \dim \operatorname{Ker}(\tilde{P}^2 | \mathcal{E}^0) - \dim \operatorname{Ker}(\tilde{P}^2 | \mathcal{E}^1).$$

When we work with infinite dimensional complex vector spaces with a Hermitian inner product, we assume that $\tilde{P}$ is a Hermitian operator.

In the case of a bounded Fredholm operator P between Hilbert spaces, Definition 1.2 and Definition 1.3 are equivalent, provided $\tilde{P}$ is introduced in the same way. We shall, mainly, adopt Definition 1.3.

1.2. A Variant of the Index 1. In general, we may regard quantities defined for a Fredholm operator P, which are independent of the choice of finite dimensional approximations, as indices in an extended sense. In the framework of functional analysis, the indices are quantities, which are invariant under continuous deformation. Here are some variants of indices.

First of all, we present "toy models" in a finite dimensional situation.

EXAMPLE 1.4. Let g^0 and g^1 be endomorphisms of complex vector spaces E^0 and E^1, respectively. If a linear mapping $f : E^0 \to E^1$ satisfies $fg^0 = g^1 f$, we have

$$\operatorname{trace}(g^0 | \operatorname{Ker} f) - \operatorname{trace}(g^1 | \operatorname{Coker} f) = \operatorname{trace}(g^0 | E^0) - \operatorname{trace}(g^1 | E^1).$$

REMARK 1.5. We can also claim the following fact related to Example 1.4. Let E^0 and E^1 be representation spaces of a compact Lie group G, and f a G-equivariant linear mapping. Then we have

$$[\operatorname{Ker} f] - [\operatorname{Coker} f] = [E^0] - [E^1]$$

as elements in the character ring $R(G)$. Here, $R(G)$ is a module consisting of formal differences between two finite dimensional representations of G. A module, obtained from a commutative semi-group by adjoining formal differences, is sometimes called the Grothendieck group.

EXAMPLE 1.6. If E carries a Euclidean inner product and f is anti-symmetric, we have

$$\dim \operatorname{Ker} f = \dim E, \qquad \operatorname{mod}\ 2.$$

REMARK 1.7. In the case of a complex vector space, we have a similar formula, provided f is anti-symmetric with respect to a non-degenerate quadratic form. Both of them are consequences of the fact that eigenvalues appear in pairs $\pm\lambda$ except the zero eigenvalue.

Both cases in examples can be generalized to infinite dimensional Fredholm setting under similar assumptions. Namely, we can define indices using finite dimensional approximation. We will get a complex valued index in the former case and $\mathbb{Z}_2 = \mathbb{Z}/2$-valued index in the latter case.

1.3. The Index for Families of Fredholm Operators. We presented an index in the case with group actions. The index was originally defined as the difference between dimensions of $\operatorname{Ker} P$ and $\operatorname{Coker} P$. If the operator P has a symmetry under a group action, we can define a quantity, which is not just about dimensions but contains information about the symmetry under the group action.

In other words, as a principle, there must be something like $\operatorname{Ker} P - \operatorname{Coker} P$, a more direct object than just the difference between dimensions, and we extract mathematically rigorous quantities and define the indices. Such an object is supposed to vary continuously, although we do not make clear what "vary continously" means.

When we want to define the "difference" between $\operatorname{Ker} P$ and $\operatorname{Coker} P$, not just as the difference between dimensions, how should we formulate it in a rigorous mathematical way?

The question here is not just about the index and a rather general question. From an object (say, P), we would like to formulate a "naive object" (e.g., the difference of $\operatorname{Ker} P$ and $\operatorname{Coker} P$) as a mathematical concept. The problem is the following: What kind of procedures and achievement realize our desire to formulate such an object?

REMARK 1.8. When a "naive object" is supposed to be a number, then there is no problem. Also, when the "procedure" is explicitly presented (like the substitution of elements in various rings into a rational polynomial), the description of the procedure itself realizes the content of the "naive object" (like morphisms between algebraic varieties defined over rational numbers).

What we discuss here is the way of defining a notion whose "character" is not clearly known in advance.

One of the answers, which is probably due to Grothendieck, is to study all continuous families of "objects" varying with parameters, not just one object (P). If a notion (corresponding to the index) is simultaneously and systematically defined for all such families, we can discuss how the index varies under deformations of P. (The meaning of "continuity" depends case by case.) This may not be the final answer, but it is certainly necessary, at least, to study families in order to construct a satisfactory theory.

Let us consider a continuous family of Fredholm operators with a parameter space. For simplicity, we assume that $\operatorname{Ker} P$ is of constant dimension for all P belonging to the family and $\operatorname{Coker} P$ is also constant dimension. Then the totality of $\operatorname{Ker} P$ with all possible P becomes a (finite dimensional) vector bundle over Z. The same thing holds for $\operatorname{Coker} P$. The point is that these vector bundles are possibly "twisted" in a topological sense. We can take the formal difference of these two vector bundles in terms of the Grothendieck group and get more information than just as a number. This is the notion of **topological K-theory**. We define the index of a family of Fredholm operators as an element in the K-group of the parameter space Z. Even in the case that $\operatorname{Ker} P$ and $\operatorname{Coker} P$ are not of constant dimension, it is known that this framework is enough to define the index. (At least, it is true for a compact topological space Z.)

The notion of the index for families of Fredholm operators is a variant of the index, but not just that.

The space of Fredholm operators (on a fixed Hilbert space) and the K-group are intimately related through the index for families. Not only is the index of families of Fredholm operators defined as elements of the K-group, but also any element in the K-group can be realized as the index of a certain family of Fredholm operators, conversely. Moreover, elements of the K-group and homotopy classes of families of Fredholm operators are in one-to-one correspondence. Shortly, the space of Fredholm operators is the classifying space for the K-group. This implies that the K-group is precisely (sufficiently and necessarily) the receiver of the index.

REMARK 1.9. The notion of the index of families can be extended, in a similar way, to the case with group actions and the case of $\mathbb{Z}_2$-valued index.

1.4. A Variant of the Index 2. Variants of the index, which we discussed so far, are topological objects in the sense that they are invariant under continuous deformations of Fredholm operators.

But there are other quantities, which depend on finer structure than topological data and are not invariant under continuous deformations.

We shall present "toy models" in a finite dimensional situation.

EXAMPLE 1.10. For a symmetric transformation f on a real vector space E equipped with a Euclidean inner product, the determinant $\det f$ is a real number determined by f.

EXAMPLE 1.11. For a symmetric transformation f on a real vector space E equipped with a Euclidean inner product, we consider the difference between the number of positive eigenvalues and the number of negative eigenvalues. This quantity is called the **signature** of f or the quadratic form associated to f.

These two quantities are related to eigenvalues of the linear mapping f. Neither of them makes sense in infinite dimensional situations without making certain "normalizations" in some sense. If we simply take the limit of the quantities for finite dimensional approximations, it diverges.

A standard way of normalization is based on the analytic continuation of so-called ζ-function, which is defined by using infinitely many eigenvalues. It is, certainly, defined under the assumption that infinitely many eigenvalues behave in a nice way. In fact, when we work with a natural set-up arising in geometric problems, such a nice behavior is guaranteed.

These quantities are the higher objects next to the index. In other words, the index is the most basic concept when dealing with these "higher objects". The index is captured by a rather rough "sieve", namely topology, but we need a finer sieve in order to capture higher objects.

In the geometry of manifolds, the corresponding object to the determinant is the analytic torsion, and the corresponding object to the signature is the η-invariant.

Under a certain situation, the analytic torsion gives a topological invariant, but it still reflects a subtler information (Reidemeister torsion) than the case of the usual index (characteristic numbers). Moreover, there is a natural object (Quillen metric), which arises naturally in trying to understand the index more geometrically than just as a number (for example, in the case of index of a family of operators).

The η-invariant arises in study of the index for an operator, on a compact manifold with boundary, with a certain boundary condition in a global nature (Atiyah-Patodi-Singer). The signature in the finite dimensional case (Example 1.11) is an integer, but the η-invariant, which is defined through normalization, is a real number, which may not be an integer.

These quantities are variants (or relatives) of the index, but they are beyond the scope of this book, which studies linear differential operators from a viewpoint of topology.

2. What Is the Atiyah-Singer Index Theorem?

2.1. Elliptic Differential Operators. In a previous section, we discussed what kind of notion the index is.

Typical examples of Fredholm operators are elliptic differential operators on manifolds.

For example, let us consider the space of rapidly decreasing functions $\mathcal{S}(X)$ on $X = \mathbb{C}$, the complex plane and an endomorphism P given by the Cauchy-Riemann equation:

$$P : f(x + \sqrt{-1}\, y) \mapsto 2\overline{\partial} f = (\partial_x + \sqrt{-1}\, \partial_y) f.$$

We study the effect of P on each wave function, after decomposing rapidly decreasing functions into wave functions with various frequencies. Using the Fourier transformation, P becomes a multiplication operator, $p_z = p_x + \sqrt{-1}\, p_y$, acting on the space of rapidly decreasing functions $\mathcal{S}(\widehat{X})$ on the dual space $\widehat{X}$ of X. Here, p_x and p_y are the real coordinates of the dual space and p_z is the complex coordinate. This multiplication operator naturally comes out from the coefficient of the operator P and is called the **principal symbol** of P.

The multiplication by p_z is almost invertible. If $p_z \neq 0$, we can make a division by p_z. Note that Fredholm operators are those which give isomorphisms after ignoring finite dimensional ambiguity. Hence, P seems a candidate of Fredholm operators.

It, however, turns out that

$$\dim \operatorname{Ker} P = 0, \quad \dim \operatorname{Coker} P = \infty,$$

and P is not a Fredholm operator. The division by p_z cannot make sense, even after ignoring finite dimensional ambiguity. In fact, we cannot divide functions in $\mathcal{S}(\widehat{X})$, which behave like polynomials in $\overline{p_z}$, by p_z.

To investigate the behavior, around $p_z = 0$, of functions on $\widehat{X}$ corresponds to the study of wave components of low frequencies, whose effects remain at infinity on the original space X. For example, we can cut off constant functions to get compactly supported functions, which contain many such wave components. Existence of such functions is an obstruction for P to being Fredholm.

Conversely, if we restrict the behavior of functions on X at infinity, we can get a Fredholm operator.

One of such realizations is to work with a torus $\mathbb{C}/L$, which is the quotient of $X = \mathbb{C}$ by a lattice $L = \mathbb{Z} + \mathbb{Z}\tau \quad (\Im\tau > 0)$. After introducing such a compact space, the Cauchy-Riemann operator becomes Fredholm. Using Fourier expansion, we can see that Ker and Coker are one-dimensional and its index is zero.

Another way is to add a term, without containing differentiation, to the Cauchy-Riemann operator. Then the behavior at infinity is automatically controlled. We shall explain an explicit way of doing this in Chapter 3.

In general, it is not difficult to see whether differential operators with constant coefficients acting on functions on an n-dimensional torus $(\mathbb{R}/2\pi\mathbb{Z})^n$ is Fredholm, using the Fourier transformation. Write the coordinates $q^1, q^2, \ldots, q^n \mod 2\pi\mathbb{Z}$. Then a homogeneous differential operator, of order k, with constant coefficients is written as $\sigma(\partial_{q^1}, \partial_{q^2}, \ldots, \partial_{q^n})$, where σ is a homogeneous polynomial of order k with constant coefficients. The Fourier transformation maps rapidly decreasing functions on X to rapidly decreasing sequences with multi integer valued indices $p_1, p_2, \ldots, p_n$. Also, the differential operator becomes a multiplying operator by $\sigma(p_1, p_2, \ldots, p_n)$. A simple sufficient condition for this operator to be Fredholm is that $\sigma(p_1, p_2, \ldots, p_n)$ never vanishes for $(p_1, p_2, \ldots, p_n) \in \mathbb{R}^n - \{0\}$.

The definition of σ above can be extended to arbitrary linear differential operator P acting on sections of vector bundles over a closed manifold and is called the **principal symbol** of P.

If the principal symbol $\sigma(P)(v)$ has an inverse for non-zero v, P is called an **elliptic differential operator**. It is known that an elliptic differential operator of order k is Fredholm as an operator acting on smooth sections of vector bundles. As we discussed in a previous section, we need to take appropriate completions of function spaces in order to apply the general framework in Banach spaces. But completions are not unique. Rather, we could say that the essential feature of P is considered as an operator acting on smooth sections and the choice of completion is a means to investigate the operator P.

Eventually, the index is determined by the homotopy class of the principal symbol of the elliptic differential operator P of order k.

The Atiyah-Singer index theorem presents the way the index is expressed.

2.2. The Dirac Operator. It may be natural to think first order differential operators are basic among all differential operators. The Cauchy-Riemann operator is one of such operators. Are there examples of other first order elliptic differential operators? The **Laplace operator** $-(\partial_{q^k})^2$ is a typical example of second order elliptic differential operators. In fact, it is the only second order linear differential operator, up to constant multiple, acting on functions, which is invariant under parallel translations and rotations in the Euclidean space. If we can find a first order linear differential operator whose square is the Laplacian, then it must be elliptic. Although there are no such operators acting on functions, there are such operators after introducing differential operators with coefficients in matrices. It was Dirac who first noticed this fact. He studied how to deal with electrons in relativity in the framework of quantum mechanics.

Let us suppose that a linear differential operator of the form $D = \sum_k \gamma^k \partial_{q^k}$ is such an operator. Here γ^k are constant matrices. Then the coefficient matrices γ^k should satisfy the following:

$$\gamma^k \gamma^l + \gamma^l \gamma^k = -2\delta^{kl},$$

where δ^{kl} is the Kronecker delta. The algebra generated by these γ^k is called the **Clifford algebra**. In short, the **Dirac operator** is a formally self-adjoint operator associated to a representation of the Clifford algebra.

We can generalize the notion of the Dirac operator on other manifolds. In this book, we discuss the Atiyah-Singer index theorem for these Dirac operators. More precisely, we deal with Dirac operators which carry a $\mathbb{Z}_2$-grading structure. We can define the index of the Dirac operator using the square of the Dirac operator as in Definition 1.3.

REMARK 1.12. To be precise, Dirac studied an object corresponding to the differential operator $\sqrt{-1}\,\partial_{q^0} + D$ on $\mathbb{R} \times X$. Here we denote by q^0 the coordinate of $\mathbb{R}$ and by D an operator of Dirac type, namely, the square of D being the Laplacian. We assume that D acts on sections of a $\mathbb{Z}_2$-graded vector bundle W and is of degree 1. The differential operator above acts on sections of a vector bundle obtained by pulling back W to $\mathbb{R} \times X$. Denote by ϵ_W the involutive automorphism corresponding to the $\mathbb{Z}_2$-grading structure. Identifying $\sqrt{-1}\,\epsilon_W$ and γ^0 corresponding to the direction of q^0, we can construct a Clifford algebra on $\mathbb{R} \times X$. Moreover, the operator

$$\widetilde{D} = \epsilon_W(\sqrt{-1}\,\partial_{q^0} + D)$$

is an operator of Dirac type on $\mathbb{R} \times X$. The notion of operators of Dirac type does not depend on the direct product decomposion of $\mathbb{R} \times X$ into $\mathbb{R}$ and X. This fact corresponds to the fact that the equation found by Dirac is "relativistic", i.e., it is independent of the choice of axes of the time ($\mathbb{R}$) and the space (X). [3]

2.3. Quantum Mechanics and Locality. The square of an operator of Dirac type is an operator of Laplace type, which resembles the Laplace operator. On the n-dimensional Euclidean space, a typical example of operators of Laplace type is given by

$$P = -\sum_{k=1}^{n} \partial_{q^k}^2 + V(q^1, q^2, \ldots, q^n).$$

Here V is a real valued function on $\mathbb{R}^n$. These operators appear in Schrödinger equations in (time independent) quantum mechanics. The first term, which is a genuine Laplace operator, corresponds to the kinetic energy in the classical theory and the second term V corresponds to the potential energy and is called the **potential**. The sum P, which corresponds to the total energy, is called the Hamiltonian operator.

REMARK 1.13. In the setting of quantum mechanics, X is the "space", which is not involved with the time. To include the "time axis", the space $\mathbb{R} \times X$ in Remark 1.12 is the "space-time".

Denote by $K(\overline{\lambda})$ the subset of X, where V is at most $\overline{\lambda}$. Let us suppose that $K(\overline{\lambda})$ is compact.

In the classical theory, particles with energy being at most $\overline{\lambda}$ move only in the bounded region $K(\overline{\lambda})$.

In quantum mechanics, we consider that states with energy being at most $\overline{\lambda}$ are represented by eigenfunctions with eigenvalues of P being at most $\overline{\lambda}$.

[3]More precisely, in the situation of Dirac's consideration, the metric on $\mathbb{R} \times X$ is not Riemannian, but the $\mathbb{R}$ direction is negative definite. So it is a Lorentzian metric. In this setting, we write $\gamma^0 = \epsilon_W$.

As we will see in Chapter 5, eigenfunctions with eigenvalue less than $\overline{\lambda}$ decay exponentially outside of $K(\overline{\lambda})$. This is a mathematical manifestation of the correspondence between the classical theory and quantum mechanics. (They are, however, not exactly zero outside of $K(\overline{\lambda})$. This is the tunneling effect.)

In other words, eigenfunctions with eigenvalue less than $\overline{\lambda}$ are almost "localized" to $K(\overline{\lambda})$.

A typical example is the case that $V = \sum_k (q^k)^2/2$, which corresponds to oscillation of a spring following the Hoock's law. In quantum mechanics, the corresponding system is called the **harmonic oscillator**.

The harmonic oscillator is not only simple in V, but also has basic significance in quantum mechanics. Because the algebraic structure (the Heisenberg algebra) behind the harmonic oscillator is used in formulation of annihilation and creation of particles in quantum field theory, in which an arbitrary number of particles can appear. In particular, we can describe annihilation and creation of n kinds of particles by using the harmonic oscillator on $\mathbb{R}^n$. It is important to note that the classical particles moving in $\mathbb{R}^n$ and the particles above, which may be annihilated and created, are completely different objects. Let us try to describe the annihilation and creation of particles in quantum field theory on a manifold X. For simplicity, we assume only one kind of particle. Then we put a copy of $\mathbb{R}$ at each point of X. Namely, we introduce a vector bundle over X with a fiber $\mathbb{R}$. Then we can consider a harmonic oscillator on the fiber over each point. The m-th eigenfunction of the harmonic oscillator on a fiber over x is regarded as the state with m particles at x. In this physical description, we need not think of the vector bundle with a fiber $\mathbb{R}$ as a geometric reality. We only need the structure of the Heisenberg algebra behind the harmonic oscillator.

2.4. Supersymmetric Harmonic Oscillator and a Proof of the Index Theorem. We can use the localizing phenomenon for eigenfunctions of operators of Laplace type as a basic tool to compute the index of the Dirac operator. In other words, the index, itself, is also "localized" in some sense.

This fact is one of reasons why the index of the Dirac operator can be written in terms of topological invariants (characteristic classes).

These invariants are related to cut-and-paste construction of vector bundles, on which the Dirac operator acts.[4]

The most basic example of operators of Dirac type is the **supersymmetric harmonic oscillator**, which is the harmonic oscillator coupled with a Clifford algebra structure, on the Euclidean space. In mathematical language, it is the conjugate of the de Rham complex (twisted by the orientation local system) by the multiplicative action by the function $\exp \sum_k (q^k)^2/2$.

We mentioned that the harmonic oscillator appears in the framework of describing the annihilation-and-creation of arbitrary number of particles. "Particles" in this context are called bosons in physics. The Clifford algebra is used in the framework of describing the annihilation-and-creation of other kinds of particles, called fermions. The supersymmetric harmonic oscillator involves an operator, which exchanges bosons and fermions (namely, a linear combination of the annihilation operator and the creation operator). This is the reason why it contains the word "supersymmetric". A mathematical expression for this symmetry among bosons and fermions is the symmetry of non-zero eigenvalues in terms of $\mathbb{Z}_2$-graded structure, which is mentioned just before Definition 1.3.

Suppose that $\mathbb{R}^n$ is assigned to each point of X. In other words, we consider a vector bundle over X with a fiber $\mathbb{R}^n$. Then, as we have seen above, we can introduce the supersymmetric harmonic oscillator on each fiber. In order to describe the annihilation-and-creation of bosons and fermions, it is not necessary to regard the vector bundle with a fiber $\mathbb{R}^n$ as a geometric reality, but we only need the structure of the Heisenberg algebra and the Clifford algebra.

However, we may ask whether this vector bundle can be considered as a geometric reality. For example, consider the situation that the manifold X is embedded in a bigger manifold $\widetilde{X}$ as a closed submanifold. Then a neighborhood of X is diffeomorphic to the total space of the normal bundle ν of X in $\widetilde{X}$. What can be deduced from understanding the family of supersymmetric harmonic oscillators acting on each fiber of ν?

The aim of this book is to explain a proof of the Atiyah-Singer index theorem from this viewpoint, in which we regard the above family as a geometric reality. Putting physical meaning and terminology

---

[4]There is another way of relating the index directly to local invariants without using potentials. Namely, it is the "heat kernel method". We, however, do not explain this argument in this book.

aside, it is just the procedure of constructing an operator of Dirac type on the total space of ν.

For a given operator of Dirac type on X, we first embed X in the Euclidean space of a sufficiently large dimension. Let $\widetilde{X}$ be this Euclidean space. The basic strategy of the proof of the index theorem given in Chapter 6 is to introduce the family of supersymmetric harmonic oscillators on ν as an auxiliary tool. This "auxiliary field" enables us to reduce our index computation to an easier case, as some computation of integration becomes easier after introducing auxiliary variables.[5] In our case, the index computation reduces to the index computation of the Dirac operator on Euclidean spaces. The last step of computing the index on Euclidean spaces can be performed with the help of the Bott periodicity theorem.

In other respects, we may regard this proof as based on the phenomenon that the index of the operator of Dirac type on the Euclidean space is localized to a neighborhood of a closed manifold X.

For a family of operators of Dirac type, we can also get the formula for the index of the family in a similar way.

2.5. Elliptic Differential Operators in Geometry. A smooth manifold is different from a topological space in the sense that the notion of differentiability of functions is assigned by the definition. Thorough discussion on smooth functions must reveal its differential topological feature. A typical example is the de Rham cohomology theory, in which topological invariants are derived using the exterior differentiation.

Some of such topological invariants are known to be given by the index of a certain elliptic differential operator.

Through the index theorem, we get information on some properties of the manifold in such a case. A typical example is the Hirzebruch signature theorem.

Associated to a geometric structure, say a complex structure, on a manifold, we get an elliptic differential operator so that its index carries some geometric meaning. Such a consideration leads us to understanding the Riemann-Roch-Hirzebruch theorem for closed complex manifolds.

---

[5] It is easier to compute the integration of $e^{-x^2-y^2}$ on $\mathbb{R}^2$ than the integration of e^{-x^2} on $\mathbb{R}$.

We can also understand the Riemann-Roch-Hirzebruch theorem for families of complex manifolds in a similar way. In fact, Grothendieck had proven the Riemann-Roch-Grothendieck theorem in the realm of algebraic geometry (for arbitrary characteristics), before the proof of the index theorem for families given by Atiyah-Singer. The notion of "family" is given in the framework of fiber bundles, but Grothendieck established the extension of the Riemann-Roch theorem for morphisms, which are not necessarily the projection of fiber bundles. Historically, it is the first time the notion of K-group emerged (in the context of algebraic geometry).

2.6. Relations with K-Theory. Let us recall that the notion of K-group appears naturally in the course of defining the index of Fredholm operators.

On the other hand, the notion of K-group provides an appropriate "expression" in each step of the proof of the index theorem mentioned above. These two aspects are not independent. We can mix them up. Or it may be more adequate to claim that there is a category of manifolds in which these aspects are better to be mixed up. Like the Riemann-Roch-Grothendieck theorem really fits the category of algebraic varieties.

To begin with, how are operators of Dirac type constructed on manifolds? Having one operator of Dirac type, we can take a tensor product with arbitrary vector bundles (in the level of principal symbols) and obtain new operators of Dirac type.

- If we have an operator of Dirac type, which can be considered as a "basic object" (or a generator), on a manifold X, then all operators of Dirac type should be obtained in this way. In such a case, the manifold X is said to carry a $Spin^c$-structure. For example, complex manifolds carry $Spin^c$-structure.
- Meanwhile, the operator of Dirac type, appearing in the Hirzebruch signature theorem, always exists for any oriented manifold of even dimension. This is not a "basic object". We, however, can consider a system of all operators of Dirac type, which are obtained by tensoring vector bundles.

In the above two categories, there is a mechanism, which provides an operator of Dirac type to arbitrary vector bundle on the manifold.

Let us suppose that we have a proper continuous mapping $\phi : X \to Y$ in each category as above. We can write ϕ as a composition of an embedding and a projection mapping. For example, identifying X with the graph of ϕ, ϕ is expressed as a composition of an embedding of X to $X \times Y$ and the projection mapping from $X \times Y$ to Y. This embedding appears in the proof of the index theorem. The projection mapping appears in study of the index for families. Taking this point into account and using the mechanism of constructing operators of Dirac type from vector bundles, we can see that these mappings induce mappings between K-groups. By composing them, we get $\phi_! : K(X) \to K(Y)$. The resulting mapping does not depend on the way of decomposing ϕ into an embedding and a projection mapping and is considered as a basic homomorphism between K-groups in each category above.

A topological description of this $\phi_!$ is the differential topological version of the Riemann-Roch-Grothendieck theorem and was carried out by Atiyah and Hirzebruch.[6]

The formulation of the index theorem using K-groups indicates that an arbitrary vector bundle over X may arise as the index of a family of something parametrized by X. As if we capture a vector bundle in reality as a finite dimensional approximation of a certain infinite dimensional object.

Thinking in this way, it is not hard to prove the Bott periodicity theorem. Atiyah realized that it naturally follows from properties of indices for families, by using the supersymmetric harmonic oscillator as an "auxiliary field".

The index for a family resembles the "integration along fibers" in an analogy with the de Rham cohomology theory. We can, actually, construct a generalized cohomology theory by extending the K-group to graded objects. Then the mapping $\phi_!$, associated to a proper continuous mapping ϕ, is understood as the Gysin mapping.

The key step in constructing a generalized cohomology theory out of K-group is the Thom isomorphism theorem for K-groups. This is an extension of the Bott periodicity theorem and proven in a similar way.

2.7. The Role and the Status of the Index of Elliptic Operators. As we have seen in a previous section, the description in terms of K-groups indicates a way of unifying a series of theories around the index theorem.

---

[6]Atiyah and Hirzebruch constructed the mapping between K-groups over $\mathbb{Q}$, more generally, i.e., without assuming existence of differential operators.

On the other hand, however, the notion of the index of elliptic linear differential operators can be seen as the first step toward the following two directions:

(1) higher invariants, which we mentioned in §1.4.
(2) study of non-linear operators.

For the former (1), it requires more than topological information as K-group, which deals with all stable homotopy classes.

As for the latter (2), there are known non-trivial examples related to geometry in dimension at most 4. In those cases, the index theorem (for families) is a basic tool, and the non-linear theory enjoys some properties similar to "locality" (Floer theory).

The notion of the index of elliptic linear differential operators carries "topological" features from the aspect (1) and is in debt to "linearity" from the aspect (2).

It is a challenging problem to define certain higher invariants, more than topological and related to non-linear differential equations, through some normalization procedure, which mixes up two aspects (1) and (2). This does not seem just one problem, and such a framework seems to be strongly requested when we try to understand some geometric phenomena in reality.

3. 1-Dimensional Case

In this section, we explain the index theorem on 1-dimensional manifolds. In this case, a solution of an ordinary linear differential equation can be uniquely determined by initial conditions. Based on this fact, the proof of the index theorem is reduced to linear algebra.

In higher dimensions, a manifold can be "curved" in various ways and we cannot hope to make an approach to the index theorem based on existence and uniqueness of the initial value problem. And we can capture the index, which is invariant under deformation, only after extending our consideration outside of genuine solutions, e.g., eigenfunctions with non-zero eigenvalues.

The argument in the 1-dimensional case depends heavily on special features in dimension 1, and it may seem that it does not mean much, although it provides examples of the index theorem.

It is, however, possible to regard a first order linear differential equation on $X = Z \times \mathbb{R}$ or $X = Z \times S^1$ (with Z a closed manifold or requiring certain appropriate conditions at infinity) as an ordinary differential equation with coefficients in matrices of infinite size.

In such cases, the argument for ordinary differential equations presents a nice model for investigation. We do not discuss these cases in detail, but explain a bit about "toy models" using ordinary differential equations.[7]

Connected 1-dimensional manifolds are $\mathbb{R}$ and S^1.

3.1. The Index Theorem on S^1. Let $A : \mathbb{R} \to M_r(\mathbb{C})$ be a smooth mapping with values in complex matrices of size r. Here, we assume that $A(x)$ is periodic with the period R: $A(x + R) = A(x)$. For $\mathbb{C}^r$-valued functions $f(x)$ and $g(x)$, we consider the following two linear differential equations:

$$\frac{df(x)}{dx} + A(x)f(x) = 0 \tag{1.1}$$

$$-\frac{dg(x)}{dx} + A(x)^*g(x) = 0 \tag{1.2}$$

Here $A^*(x)$ is the adjoint matrix of $A(x)$. We, in practice, study differential equations on $S^1 = [0, R]/(0 \sim R)$.

If we do not require the periodicity condition, the solution uniquely exists for a given initial value at a point. Hence the dimension of the space of periodic solutions is at most r. According to changes of $A(x)$, the dimension of periodic solutions varies and takes values between 0 and r.

The index theorem, in this setting, is the following:

THEOREM 1.14 (The index theorem on S^1). *The dimensions of these two differential equations coincide.*

Before the proof, we present a condition for a solution to be periodic. We abbreviate the first equation in the form $(\nabla + A)f = 0$. Solving this equation on $[0, R]$, we get a linear mapping $\phi : \mathbb{C}^r \to \mathbb{C}^r$, which assigns the value $f(R) = v_1$ at $x = R$ of the solution to an arbitrary initial value $f(0) = v_0$ at $x = 0$. The periodic solutions are exactly those corresponding to fixed vectors of ϕ.

In the same way, we get a linear mapping $\psi : \mathbb{C}^r \to \mathbb{C}^r$ from the second equation $(-\nabla + A^*)g = 0$. The periodic solutions correspond to fixed points of ψ.

---

[7]In one dimension higher, Tomoyoshi Yoshida constructed a theory, whose toy model is differential equations on $S^1 \times \mathbb{R}$ or $\mathbb{R} \times \mathbb{R}$. He used it in the study of Floer homology.

Therefore the theorem follows from the fact that ϕ and ψ^{-1} are adjoint, which we see below. We have to note certain relations between two differential equations for the proof. Namely, $\nabla + A$ and $-\nabla + A^*$ are "formally adjoint" to each other.

We review formal adjoint operators.

We define a Hermitian inner product on $\mathbb{C}^r$ by

$$\langle u, v\rangle = \sum_{i=1}^{r} u_i \overline{v}_i,$$

for $u = (u_1, u_2, \ldots, u_r)$ and $v = (v_1, v_2, \ldots, v_r)$. Then the fact that the operators $\nabla + A$ and $-\nabla + A^*$ are formally adjoint means that the following equality holds:

$$\int_{x_0}^{x_1} \langle(\nabla + A)f, g\rangle dx - \int_{x_0}^{x_1} \langle f, (-\nabla + A^*)g\rangle dx = \langle f(x_1), g(x_1)\rangle - \langle f(x_0), g(x_0)\rangle.$$

If f or g vanishes at $x = x_0, x_1$, the right hand side becomes zero. Or, as we assume right now, if f and g are periodic functions with the period R, then we have

$$\int_0^R \langle(\nabla + A)f, g\rangle dx = \int_0^R \langle f, (-\nabla + A^*)g\rangle dx.$$

From these equalities, we formally derive that $\nabla + A$ and $-\nabla + A^*$ are adjoint with respect to the Hermitian inner product:

$$(f_1, f_2) = \int_0^R \langle f_1, f_2\rangle dx, \quad (g_1, g_2) = \int_0^R \langle g_1, g_2\rangle dx.$$

Now we give a proof of the index on S^1.

PROOF. If functions f and g, with values in $\mathbb{C}^r$, on $\mathbb{R}$ satisfies $(\nabla + A)f = 0$ and $(-\nabla + A^*)g = 0$, we have

$$\langle f(x_1), g(x_1)\rangle = \langle f(x_0), g(x_0)\rangle.$$

Putting $x_0 = 0$ and $x_1 = R$, it implies that ϕ and ψ^{-1} are adjoint. $\square$

3.2. The Index Theorem on $\mathbb{R}$. Next, we give a similar discussion on differential equations over $\mathbb{R}$. Solutions here are supposed to converge to zero at infinity. The goal is to compare the dimensions of the spaces of solutions of $(\nabla + A)f = 0$ and $(-\nabla + A^*)g = 0$, respectively. Then we are required to study, closely, their behavior at infinity.

Let $A : \mathbb{R} \to M_r(\mathbb{C})$ be a smooth mapping with values in complex square matrices of size r, which satisfies

$$A(x) = \begin{cases} A_0 & (x \leq R_0) \\ A_1 & (x \geq R_1) \end{cases}$$

for some complex matrices A_0 and A_1. We consider the following ordinary differential equation for a vector valued function $f(x)$, i.e., $f : \mathbb{R} \to \mathbb{C}^r$,

$$\frac{df(x)}{dx} + A(x)f(x) = 0.$$

Pick a point on $\mathbb{R}$. For any initial value, the solution exists uniquely. Hence the space of solutions is an r-dimensional complex vector space. We, however, restrict ourselves to solutions which decay at infinity.

We write the space of such solutions

$$\operatorname{Ker}(\nabla + A) = \left\{ f(x) \;\middle|\; \frac{df(x)}{dx} + A(x)f(x) = 0,\ \lim_{x\to\pm\infty} f(x) = 0 \right\}.$$

Here, we use the notations $\nabla = \frac{d}{dx}$, $A = A(x)$. Changing the function $A(x)$ on a compact interval $-R \leq x \leq R$, the dimension of this vector space may vary.

Now, we introduce another ordinary differential equation

$$-\frac{dg(x)}{dx} + A^*(x)g(x) = 0.$$

Here, $g(x)$ is a $\mathbb{C}^r$-valued function. Write

$$\operatorname{Ker}(-\nabla + A^*) = \left\{ g(x) \;\middle|\; -\frac{dg(x)}{dx} + A^*(x)g(x) = 0,\ \lim_{x\to\pm\infty} g(x) = 0 \right\}.$$

The dimension of this vector space may also vary according to changes of $A^*(x)$, in other words, changes of $A(x)$.

A remarkable fact is that the difference between the dimensions of $\operatorname{Ker}(\nabla + A)$ and $\operatorname{Ker}(-\nabla + A^*)$ does not change, even though $A(x)$ varies. More precisely, we have the following:

THEOREM 1.15 (the index theorem on $\mathbb{R}$). *If 0 is not the real part of eigenvalues of A_0 and A_1, we have*

$$\dim \operatorname{Ker}(\nabla + A) - \dim \operatorname{Ker}(-\nabla + A^*) = \frac{\operatorname{sign}(A_1)}{2} - \frac{\operatorname{sign}(A_0)}{2}.$$

Here $\operatorname{sign}(B)$ *for a square matrix B denotes the number of eigenvalues of B with positive real part subtracted by the number of those with negative real part.*

Note that the right hand side only depends on A_0 and A_1. The left hand side, the difference of dimensions of solution spaces, is the index of the operator $\nabla + A$.

Since the proof is simple, we omit it.

REMARK 1.16. As we noticed before, arguments for ordinary differential equations can be a model for a part of the argument for first order differential equations on a manifold $X = Z \times \mathbb{R}$. The above equality involving with sign is a "toy model" of the Atiyah-Patodi-Singer theory, which expresses the index with the help of η-invariant. More explanation on this topic is beyond the scope of this book.

3.3. The Index Theorem on $\mathbb{R}$ with Actions. Fix a Hermitian matrix T of size r. Let $A(x)$ be a smooth mapping, with values in complex matrices of size r, on $\mathbb{R}$. We assume that $A(x)$ takes constant values A_0 and A_1 on $x \leq R_0$ and $x \geq R_1$, respectively, and that 0 is not the real part of any eigenvalue of these matrices. We also require that $TA(-x) = -A(x)T$ holds, in particular, we have $TA_0 = -A_1T$. Denote by τ_V and τ_W the linear mappings $f(x) \mapsto Tf(-x)$ and $g(x) \mapsto -T^*g(-x)$, respectively. Since $T = T^*$, we have $\tau_W(\nabla + A) = (\nabla + A)\tau_V$ and $\tau_V(-\nabla + A^*) = (-\nabla + A^*)\tau_W$. Write $\mathrm{Ker}(\nabla + A)$ and $\mathrm{Ker}(-\nabla + A^*)$ the spaces of solutions which converge zero at infinity. These solution spaces are preserved under the action of τ_V and τ_W.

THEOREM 1.17 (the index theorem on $\mathbb{R}$ with actions).

$$\mathrm{trace}(\tau_V | \mathrm{Ker}(\nabla + A)) - \mathrm{trace}(\tau_W | \mathrm{Ker}(-\nabla + A^*)) = \mathrm{trace}\, T.$$

A similar result also holds on S^1. The details are left to readers.

EXERCISE 1. Formulate the index theorem on S^1 with actions and prove it. (The answer is not unique.)

3.4. The mod 2 Index Theorem in Dimension 1. Let $A(x)$ be a smooth mapping with values in skew symmetric complex matrices of size r. We consider a differential equation $(\nabla + A)f = 0$ for a $\mathbb{C}^r$-valued function $f(x)$.

(1) The case over S^1. Fix $R > 0$ and a complex orthogonal matrix O of size r. We assume twisted periodicity for $A(x)$ and $f(x)$:

$$A(x + R) = OA(x)O^{-1}, \quad f(x + R) = Of(x),$$

and consider $f(x)$ as a section of a vector bundle, twisted by O, over $S^1 = \mathbb{R}/R\mathbb{Z}$.

(2) The case over $\mathbb{R}$. Fix two real numbers $R_0 < R_1$ and two skew symmetric matrices A_0 and A_1. Here we assume that 0 is not the real part of any eigenvalue of A_0 and A_1, which only exist in the case that r is odd.

We also require that $A(x)$ takes constant values A_0 and A_1 on $x \leq R_0$ and $x \geq R_1$, respectively, and that $f(x)$ converges to 0 when $|x| \to \infty$.

Under these situations, we have the following result. (The proof is also reduced to linear algebra as discussed in a previous section.)

THEOREM 1.18.

(1). *The case over* S^1. *Write* $\epsilon = 0, 1$ *according to* $\det O = (-1)^\epsilon$. *We have*

$$\dim \operatorname{Ker}(\nabla + A) \equiv r + \epsilon \pmod 2.$$

(2). *The case over* $\mathbb{R}$. *Write* $\epsilon = 0, 1$ *according to*

$$\frac{\text{the product of eigenvalues of } A_0 \text{ with positive real part}}{\operatorname{Pf}(A_0)}$$

$$= (-1)^\epsilon \frac{\text{the product of eigenvalues of } A_1 \text{ with positive real part}}{\operatorname{Pf}(A_1)}.$$

Here Pf *denotes the Pfaffian of a skew symmetric matrix. Then we have*

$$\dim \operatorname{Ker}(\nabla + A) \equiv \epsilon \pmod 2.$$

(The usual integer valued index, defined before, is zero in these cases.)

REMARK 1.19. We give a supplementary explanation on skew symmetric matrices. For a skew symmetric matrix C, the square of $\operatorname{Pf}(C)$ equals $\det(C)$. On the other hand, if 0 is not the real part of any eigenvalue of C, the square of the product of eigenvalues with positive real part equals $(-1)^{r/2} \det(C)$. Hence the ratio of them, before taking the square, is one of $\pm(\sqrt{-1}\,)^{r/2}$. If we take a conjugation of C by a complex orthogonal matrix O, this ratio is multiplied by $\det O$. This fact implies that the set of skew symmetric matrices splits into two classes.[8]

EXERCISE 2. Prove Theorem 1.18.

[8]It is related to the way two maximally isotropic subspaces intersect.

SUMMARY.

1.1 The index of Fredholm operators is defined as the difference of the dimensions of Ker and Coker. It is also possible to define the index for Fredholm Hermitian transformations, of degree 1, on a $\mathbb{Z}_2$-graded vector space.

1.2 The index is invariant under deformations. There are variants of the index, which is invariant under deformations. In particular, we can define the index for a family of Fredholm operators, which takes values in K-group.

1.3 Elliptic linear differential operators on (closed) manifolds are Fredholm, and their indices enjoy a certain property of locality.

1.4 We can interpret the locality above using ideas in quantum mechanics. Conversely, we may put it that some arguments in quantum mechanics are justified mathematically.

CHAPTER 2

Manifolds, Vector Bundles and Elliptic Complexes

Although we assume, in this book, that readers have a basic background in the theory of manifolds and vector bundles, we summarize here what is particularly needed in later chapters.

In §1, we review the integration of differential forms. It is not used until Chapter 7, but it contains some propositions which are comparable to some consideration on the index in Chapter 6. So we start with it in order to clarify the argument. The meaning of "propositions comparable to those for the index" becomes clear when we introduce the notion of K-groups in Chapter 9 of the sequel.

In §2, we explain that any manifolds can be embedded into some infinite dimensional Euclidean space and that any vector bundles can be embedded into some trivial vector bundle of infinite rank. By taking approximation, we get an embedding into some finite dimensional Euclidean space or some trivial vector bundle of finite rank, respectively. The former, in particular, gives a procedure in the proof of the index theorem.

In §3, the notion of Clifford modules is introduced. Elliptic differential operators, which appear in this book, are operators of Dirac type, which is defined on Clifford modules. In §4, we give examples of geometric invariants, which arise in relation to operators of Dirac type.

1. Differential Forms with Compact Support and Their Integration

Unless otherwise stated, in this book manifolds are assumed to be differentiable manifolds, fiber bundles over manifolds are smooth fiber bundles, and sections of a fiber bundle are smooth.

Denote by TX the tangent bundle of an n-dimensional manifold X, and by T^*X its cotangent bundle. A differential k-form is defined as a section of the k-th exterior product bundle $\bigwedge^k T^*X$. Denote by $\Omega^k(X)$ the set of all differential k-forms on X.

Take the direct sum over all degrees, and write

$$\Omega^\bullet(X) = \sum_{k=0}^{n} \Omega^k(X).$$

The exterior differentiation d operates on $\Omega^\bullet(X)$ and increases the degree by one. In other words, d is a **operator of degree 1**.

Now we consider the "integration over X". For this, we introduce "differential forms with compact support" and the "local system of orientation".

The **support** supp ω of a differential form ω is the closure of the set of all points where ω does not vanish. Denote by $\Omega_c^k(X)$ the set of all differential k-forms with compact support, and by $\Omega_C^\bullet(X)$ the direct sum over k. The exterior differentiation d preserves $\Omega_C^\bullet(X)$.

A **local system** is a vector bundle, whose transition functions are locally constant. Namely, a vector bundle θ, of rank r, over X is called a local system, if it is given by a covering $\{U_\alpha\}$ of X and transition functions

$$g_{\alpha\beta} : U_\alpha \cap U_\beta \to GL_r(\mathbb{R}),$$

and all $g_{\alpha\beta}$ are locally constant functions. Here, $g_{\alpha\beta}$ should satisfy that $g_{\alpha\alpha} = I_r$ on U_α and that $g_{\alpha\beta}g_{\beta\gamma}g_{\gamma\alpha} = I_r$ on $U_\alpha \cap U_\beta \cap U_\gamma$. A differential k-form with coefficients in the local system θ is a section of $\theta \otimes \bigwedge^k T^*X$ and we denote by $\Omega^k(X, \theta)$ the totality of those. The exterior differentiation d naturally extends to $\Omega^\bullet(X, \theta)$.

A typical example of local systems is the local system of orientation. For a given open covering $\{U_\alpha\}$ and coordinates $f_\alpha : U_\alpha \to \mathbb{R}^n$, we define locally constant functions with values in $\{\pm 1\}$ according to the sign of Jacobian matrices of $f_\alpha \circ f_\beta^{-1}$ being positive or negative. In this way, we get a local system of rank 1, which is the local system of orientation. We denote it by θ_X in this book.

In this set-up, we have the integration over X

$$\int_X : \Omega_c^n(X, \theta_X) \to \mathbb{R}.$$

We slightly generalize the integration framework.

Let $\pi_F : F \to X$ be a fiber bundle over X. Take the local system of orientation on each fiber and consider the totality over all fibers. Then we get, naturally, a local system on F. We call it the "local system of orientation along fibers" and write it θ_{π_F}. A differential form with fiberwisely compact support is a differential form on F, so that the intersection of its support and each fiber is compact. Denote by $\Omega^\bullet_{\pi_F c}(F)$ the totality of such forms. We can also define analogous notions for differential forms with coefficients in a local system.

In this setting, we define, by the "integration along fibers", the following linear mapping, which decreases the degree by the dimension of the fiber

$$\int_{\pi_F} : \Omega^\bullet_{\pi_F c}(F, \theta_{\pi_F}) \to \Omega^\bullet(X).$$

For a local system θ on X, we also have

$$\int_{\pi_F} : \Omega^\bullet_{\pi_F c}(F, \theta_{\pi_F} \otimes \pi_F^* \theta) \to \Omega^\bullet(X, \theta).$$

Particularly, in the case that θ is the local system θ_X of orientation on X, we have

$$\theta_{\pi_F} \otimes \pi^* \theta_X \cong \theta_F,$$

and, hence,

$$\int_{\pi_F} : \Omega^\bullet_{\pi_F c}(F, \theta_F) \to \Omega^\bullet(X, \theta_X).$$

For a differential form on F with compact support, its integration along fibers is a differential form on X with compact support. The composition

$$\Omega^\bullet_c(F, \theta_F) \xrightarrow{\int_{\pi_F}} \Omega^\bullet_c(X, \theta_X) \xrightarrow{\int_X} \mathbb{R}$$

coincides with the integration over F. Rather, in logical order, we define the integration along fibers as an operation so that it satisfies this property (even when restricted to arbitrary submanifolds in X). Existence and uniqueness are shown after that.

Note that there are two basic operations for differential forms with compact support.

REMARK 2.1. In the proof of the index theorem in Chapter 6, some operations similar to these two operations turn out to be crucial.

The first operation is the following. Denote by θ_U the restriction of a local system on X to an open subset U of X. Extending differential forms by 0 outside of U, we have a natural mapping

$$i_! : \Omega^\bullet_c(U, \theta|U) \to \Omega^\bullet_c(X, \theta).$$

Here i is the embedding of U to X. In particular, when θ is the local system of orientation on X, this operation commutes with taking the integration.

The second is the exterior product as below.

The product of differential forms, with non-compact support, can be compactly supported. The situation we encounter in this book is the following. Let $\pi_F : F \to X$ be a fiber bundle. Then elements in the image of the pull-back $\pi_F^* : \Omega_c^\bullet(X) \to \Omega^\bullet(F)$ may not be compactly supported. However, their exterior product with elements in $\Omega_{\pi_F c}^\bullet(F)$ are compactly supported.

In other words, by the pull-back and the exterior product, we have

$$\Omega_{\pi_F c}^\bullet(F) \times \Omega_c^\bullet(X) \to \Omega_c^\bullet(F).$$

In the same manner, there is an operation

$$\Omega_{\pi_F c}^\bullet(F, \theta_{\pi_F}) \times \Omega_c^\bullet(X, \theta_X) \to \Omega_c^\bullet(F, \theta_F).$$

With respect to this product, the following equality holds:

$$\int_{\pi_F} \omega_1 \wedge (\pi_F^* \omega_0) = \left(\int_{\pi_F} \omega_1 \right) \wedge \omega_0,$$

where ω_0 and ω_1 are elements in $\Omega_c^\bullet(X, \theta_X)$ and $\Omega_{\pi_F c}^\bullet(F, \theta_{\pi_F})$.

2. Embeddings of Manifolds and Vector Bundles to Trivial Objects

Any manifold X can be embedded to a Euclidean space of sufficiently large dimension. Here is an argument.

To be formal, the idea is:

(1) construct an embedding of X to an infinite dimensional Euclidean space

and then

(2) approximate this embedding by a mapping to a certain finite dimensional Euclidean space, which gives rise to a desired embedding. In fact, we can construct an embedding without explicit use of an infinite dimensional Euclidean space.

Denote by $C^\infty(X)$ the set of all smooth functions on X. The idea in (1) is to regard all of them as coordinate functions. Denote by $\mathrm{Map}(C^\infty(X), \mathbb{R})$ the direct product of copies of $\mathbb{R}$ over all elements of $C^\infty(X)$. This is a really huge space. Then we get a natural mapping

$$\iota : X \to \mathrm{Map}(C^\infty(X), \mathbb{R}).$$

To study an infinite dimensional manifold seriously, we need much preliminary work, such as putting a topology on it. Here, we just explain the idea of the construction, and we do not need it. We only use the fact that an analog of the differential for ι

$$d\iota : TX \to \mathrm{Map}(C^\infty(X), T\mathbb{R})$$

is defined by taking the differential of each "coordinate".

We give a formal argument for the fact that the natural mapping ι above is an embedding. Namely, we explain

(1) (immersiveness) The mapping $d\iota$ maps any non-zero tangent vector v on X to a non-zero vector.
(2) (injectivity) The mapping ι maps distinct points x_0 and x_1 to distinct points.

Firstly, about the condition (1) for immersiveness. For a non-zero tangent vector v, there is a smooth function g so that the derivative of g in the direction of v is not zero. This implies that the coordinate function, corresponding to g, of $d\iota(v)$ is non-zero.

Secondly, about the condition (2) for injectivity. For a pair of distinct points x_0 and x_1, there is a smooth function f so that the values $f(x_0)$ and $f(x_1)$ are distinct. This implies that the points $\iota(x_0)$ and $\iota(x_1)$ take different values of the coordinate function corresponding to f.

Now we show, following the above idea, that there exists an embedding of a closed manifold to a finite dimensional Euclidean space.

THEOREM 2.2. *Let X be a closed manifold. Then there is an embedding of X to a Euclidean space of sufficiently large dimension.*

PROOF. Firstly, we discuss the condition to be an immersion. Denote by $\mathbb{P}(TX)$ the set of all possible tangent directions over all points on X and call its element a "tangent line". The space $\mathbb{P}(TX)$ is a fiber bundle over the base space X. Its fiber is a real projective space, which is compact. Hence, the total space is also compact. For any "tangent line" l at any point x, we can take a smooth function f_l, whose derivative in the direction of l is not zero. Then there is an open neighborhood U_l of l in $\mathbb{P}(TX)$ such that all derivatives of f_l in the direction in U_l are not zero. Based on compactness of $\mathbb{P}(TX)$, we cover it with finitely many open subsets $U_{l_1}, U_{l_2}, \dots, U_{l_s}$. Then the mapping $F = (f_{l_1}, f_{l_2}, \dots, f_{l_s})$ gives an immersion.

Next, we discuss the injectivity. Since the mapping F above is an immersion, its restriction to a sufficiently small neighborhood U_x

of arbitrary point x in X is injective. The complement K of the open subset $\cup_{x \in X} U_x \times U_x$ in $X \times X$ is compact. For each point $y = (x_0, x_1)$ in K, x_0 and x_1 are distinct. So there is a smooth function f_y on X such that $f(x_0)$ and $f(x_1)$ are distinct. Then there is an open neighborhood U_u of y in $X \times X$ such that $f_y(x'_0)$ and $f_y(x'_1)$ are distinct for $(x'_0, x'_1) \in U_y$. We take an open covering of the compact subset K by finitely many such open subsets $U_{y_1}, U_{y_2}, \ldots, U_{y_u}$. Now we put $G = (f_{y_1}, f_{y_2}, \ldots, f_{y_u})$. Then the mapping $(F, G) : X \to \mathbb{R}^{s+u}$ gives a desired embedding. □

EXERCISE 1. Suppose that a manifold X is covered by finitely many open subsets and that each open subset is equipped with local coordinate functions. Construct an explicit embedding of X to a (finite dimensional) Euclidean space using these local coordinates and a partition of the unity.

REMARK 2.3.

(1) In the explanation above, we used $\mathrm{Map}(C^\infty(X), \mathbb{R})$. Taking into account that the space $C^\infty(X)$ is a vector space, we can also use a (bit) smaller space $\mathrm{Hom}(C^\infty(X), \mathbb{R})$.

(2) Suppose that a complex line bundle L on X is given. We can use the set $\Gamma(L)$ of smooth sections of L, instead of the space of all functions $C^\infty(X)$ on X, and make a similar construction. A section of L does not produce a coordinate function. However, their "ratio" makes sense, when we consider several sections simultaneously. In this way, we get an embedding to an infinite dimensional complex projective space $\mathbb{P}(\mathrm{Hom}(\Gamma(L), \mathbb{C}))$. After taking a finite dimensional approximation, we obtain an embedding to a finite dimensional complex projective space. It was Kunihiko Kodaira who made such a construction for complex manifolds and studied the condition that L admits "very ample" holomorphic sections.

(3) Using this embedding, the Riemann-Roch theorem can be extended and proved for complex manifolds, which can be embedded to complex projective spaces (the Riemann-Roch-Hirzebruch theorem or the Riemann-Roch-Grothendieck theorem). The proof, as we explain in this book, is considered as a differential topological version of such an argument.

It is known that manifolds, which are not necessarily closed manifolds, can be embedded to a finite dimensional Euclidean space. What we need in this book is the following.

COROLLARY 2.4. *If X is the interior of a compact manifold $\overline{X}$ with boundary, X admits an embedding, as a closed submanifold, into a finite dimensional Euclidean space.*

PROOF. Denote by $\widetilde{X}$ the manifold obtained by gluing $\overline{X}$ and its copy along the boundary (it is called the **double** of $\overline{X}$). Let $f_0 : \widehat{X} \to \mathbb{R}^m$ be an embedding of $\widehat{X}$ to a finite dimensional Euclidean space. Choose a smooth non-negative function f on $\overline{X}$ such that f takes the value 0 exactly on the boundary. Then the mapping $(f_0|X, f^{-1}) : X \to \mathbb{R}^{m+1}$ is an embedding. □

We have an analogous theorem for vector bundles. If transition functions of a vector bundle are continuous, we call such a vector bundle a continuous vector bundle. On a topological space, the notion of "smoothness" is not defined, and we can only define continuous vector bundles.

THEOREM 2.5. *For a continuous vector bundle F on a compact topological space K, there is a continuous vector bundle G such that the direct sum $F \oplus G$ becomes isomorphic to a trivial vector bundle.*

PROOF. It is enough to show existence of a surjective homomorphism, as vector bundles, $f : K \times \mathbb{C}^k \to F$. (Then we can take $G = \operatorname{Ker} f$.) We use the same idea as in the proof of Theorem 2.2. Namely, choose sufficiently many sections of F and denote them by $s_1, s_2, \ldots, s_k$. The assertion follows from the fact that the homomorphism $f(t_1, t_2, \ldots, t_k) = t_1 s_1 + t_2 s_2 + \cdots + t_k s_k$ is surjective. It is not difficult to prove it, using the local triviality of F and the partition of the unity. The details are left to readers. □

We denote by $Gr_r(\mathbb{C}^N)$ the set of all r-dimensional subspaces in $\mathbb{C}^N$ and call it the Grassmannian manifold. It, naturally, carries a structure of a manifold and a smooth vector bundle $F_{r,N}$ of rank r. Theorem 2.5 can be rephrased as

PROPOSITION 2.6. *For a continuous vector bundle F, of rank r, on a compact topological space K, there is a continuous mapping from K to $Gr_r(\mathbb{C}^N)$, for a sufficiently large N, such that F is isomorphic to the pull-back of $F_{r,N}$.*

A natural embedding $\mathbb{C}^{N_0} \to \mathbb{C}^{N_1}$, for $N_0 < N_1$, induces a natural embedding $Gr_r(\mathbb{C}^{N_0}) \to Gr_r(\mathbb{C}^{N_1})$. We denote by $Gr_r(\mathbb{C}^\infty)$ its inductive limit, on which a natural vector bundle $F_{r,\infty}$ exists. A continuous vector bundle over a compact space is isomorphic to the pull-back of the vector bundle $F_{r,\infty}$ by some continuous mapping to $Gr_r(\mathbb{C}^\infty)$. It is also known that the homotopy class of the continuous mapping above is uniquely determined (the proof is not difficult, either).

DEFINITION 2.7. Write $BU(r) = Gr_r(\mathbb{C}^\infty)$ and call it the **classifying space** of $U(r)$. The isomorphism classes of continuous vector bundle, of rank r, on a compact space are classified by homotopy classes of continuous mappings to $BU(r)$.

Let X be a manifold, which has the same homotopy type as a compact space. Suppose that a continuous vector bundle is given on X. Then, as we saw in Propositon 2.6, it is isomorphic to the pull-back of $F_{r,N}$ by some continuous mapping from X to $Gr_r(\mathbb{C}^N)$. A continuous mapping between manifolds is approximated by a smooth mapping, and two such approximations can be joined by a smooth homotopy. The pull-back of $F_{r,N}$ by a smooth mapping from X to $Gr_r(\mathbb{C}^N)$ is smooth, and this fact implies the following

LEMMA 2.8. *A continuous vector bundle over a manifold, of the same homotopy type as a compact space, carries a structure of a smooth vector bundle, which is unique (up to homotopy).*

Therefore, it is not necessary to distinguish whether a vector bundle is defined in the continuous category or the smooth category.

3. Clifford Modules and Operators of Dirac Type

3.1. Clifford Algebra and Spinor Representations. Let E be an n-dimensional real vector space with a Euclidean metric.

DEFINITION 2.9. The Clifford algebra $Cl(E)$ over E is the algebra which is generated by E and satisfies the following relations:

$$uv + vu = -2(u, v),$$

where u, v are elements in E and (u, v) is the inner product of u and v.

In particular, we have $v^2 = -|v|^2$. Choose an orthonormal basis $e_1, e_2, \ldots, e_n$ of E. Then $\{e_1^{i_1} e_2^{i_2} \cdots e_n^{i_n}\}$, where $i_1, i_2, \ldots, i_n = 0, 1,$

gives a basis of $Cl(E)$ and its dimension is 2^n. Particularly, the element vol, which is defined by

$$\mathrm{vol} = e_1 e_2 \cdots e_n,$$

is invariant under the change of a basis by the action of $SO(n)$ and only depends on the orientation of E. Reversing the orientation, the sign **vol** is reversed. If the orientation is not specified, **vol** is the element, unique up to sign, in $Cl(E)$.

Noting that the algebra $Cl(E)$ consists of linear combinations of products of several elements of E, we define $Cl^0(E)$ (resp. $Cl^1(E)$) the totality of the linear combinations of products of even (resp. odd) numbers of elements of E. The intersection of these two subspaces is $\{0\}$, and we have the direct sum decomposition

$$Cl(E) = Cl^0(E) \oplus Cl^1(E).$$

Regarding the superscript 0,1 as elements in $\mathbb{Z}_2$, which we call the **degree**, the product in $Cl(E)$ respects the degree as follows:

$$Cl^i(E) \otimes Cl^j(E) \to Cl^{i+j}(E).$$

Namely, $Cl(E)$ carries a structure of a $\mathbb{Z}_2$-graded $\mathbb{R}$-algebra.

3.1.1. *Unitary representations of the Clifford algebra.* Because of the minus sign in the definition of the Clifford algebra, the following definition makes sense.

DEFINITION 2.10. A unitary representation of the Clifford algebra $Cl(E)$ is a pair (R, c_R) of a complex vector space R and a real-linear mapping $c_R : E \to \mathrm{End}_{\mathbb{C}} R$ satisfying the following conditions.

- c_R extends to a homomorphism $Cl(E) \to \mathrm{End}_{\mathbb{C}} R$ as $\mathbb{R}$-algebras.
- For each element v in E, $c_R(v)$ is skew-hermitian.

The mapping c_R is called the Clifford multiplication on R. We may also denote by c_R the $\mathbb{R}$-algebra homomorphism $Cl(E) \to \mathrm{End}_{\mathbb{C}} R$.

A unitary representation is called irreducible, if it does not contain non-trivial sub-representation.

THEOREM 2.11.

(1) *If $n = 2m$, $Cl(E)$ has a unique irreducible unitary representation of dimension 2^m.*

(2) *If* $n = 2m + 1$, $Cl(E)$ *has exactly two irreducible unitary representations, each of which is of dimension* 2^m. *These two representations are distinguished by the action of* vol *being* $\pm(\sqrt{-1}\,)^{m+1}$. *(In other words, there are two representations according to orientations of* E.*)*

PROOF. We only give a sketch of the proof. Denote by G the finite group, of order 2^{n+1}, generated by $\{\pm e_1^{i_1} e_2^{i_2} \cdots e_n^{i_n}\}$. Unitary representations of $Cl(E)$ are exactly equal to unitary representations of G such that $-1 \in G$ acts by the multiplication by -1. Thus the problem is reduced to classification of unitary representations of G. If n is even, the center of G equals $\{\pm 1\}$. If $n = 2m + 1$ is odd, the center of G equals $\{\pm 1, \pm \mathrm{vol}\}$, and the equality $\mathrm{vol}^2 = (-1)^{m+1}$ holds. If an element $e \in G$ does not belong to the center, all elements conjugate to e are $\pm e$. Hence, the number of conjugacy classes is $(2^{n+1} - 2)/2 + 2 = 2^n + 1$, if n is even, and $(2^{n+1} - 4)/2 + 4 = 2^n + 2$, if n is odd. Note that the number of conjugacy classes equals the number of irreducible unitary representations. Since the group $G/\{\pm 1\}$ is a commutative group of order 2^n, there are 2^n 1-dimensional irreducible representations of G such that -1 acts trivially. Hence the number of unitary representations of $Cl(E)$ is one, if n is even and two, if n is odd. When n is odd, the outer automorphism of G, induced by $e_i \mapsto -e_i$, interchanges two unitary representations. Thus they have the same dimension and vol acts on them differently by sign. Meanwhile, the sum of dimensions of all irreducible representations coincides with the order of G. The dimension of irreducible representations is determined from this fact. □

If the vector space E is a subspace of a bigger vector space $\widetilde{E}$ equipped with a Euclidean metric, a unitary representation space $\widetilde{R}$ of $Cl(\widetilde{E})$ is also considered as a unitary representation space of $Cl(E)$ by restricting the action. Such a representation carries an additional structure, which comes from the structure of $Cl(\widetilde{E})$, which we explain next. The "additional structure" is considered as $H(R)$ in the following definition.

DEFINITION 2.12. Denote by $\mathrm{Herm}\, R$ the set of all Hermitian transformations on a unitary representation space R of $Cl(E)$ and by $H(R)$ the set of all element h of $\mathrm{Herm}\, R$, which anti-commute with $c_R(v)$ for all v in E: $c_R(v)h + hc_R(v) = 0$.

Let E' be the orthonormal complement of E in $\widetilde{E}$. Then we have

- For an element v' in E', $\sqrt{-1}\, c_{\widetilde{R}}(v')$ belongs to $H(R)$.
- In particular, when $E' = \mathbb{R}e_0$, the element $\epsilon_R = \sqrt{-1}\, c_{\widetilde{R}}(e_0)$ is Hermitian and unitary, and satisfies $\epsilon_R^2 = 1$.

Namely, by knowing how $H(R)$ is big, we can, roughly, see whether a given unitary representation R of $Cl(E)$ is considered as the restriction of a unitary representation of $Cl(\widetilde{E})$, for a strictly larger space $\widetilde{E}$ than E.

When $\widetilde{E} = \mathbb{R}e_0 \oplus E$ and a representation R is obtained by restricting a unitary representation of $Cl(\widetilde{E})$, the eigenspace decomposition under the action of ϵ_R induces a $\mathbb{Z}_2$-graded structure on R, which is compatible with the Clifford multiplication.

We will use, later, this $\mathbb{Z}_2$-graded unitary representation in relation with the formulation of the index theorem.

First of all, we review the notion of $\mathbb{Z}_2$-graded vector spaces.

DEFINITION 2.13.

(1) A vector space E is a $\mathbb{Z}_2$-graded vector space, if assigned is an element ϵ_E of $\operatorname{End} E$ satisfying $\epsilon_E^2 = 1$. Denote by E^0 and E^1 the eigenspaces of ϵ_E with eigenvalues 1 and -1, respectively. Then the assignment of ϵ_E is equivalent to the direct sum decomposition $E = E^0 \oplus E^1$. We call the superscript $0, 1$ the degree. For an element a of E^k, we write the degree of a by $k = \deg a$.

(2) A vector space E is a $\mathbb{Z}_2$-graded Hermitian vector space, if assigned is the Hermitian metric on E, which is invariant under the action of ϵ_E. In this case, ϵ_E is unitary and Hermitian. This condition is equivalent to assigning Hermitian metrics on E^0 and E^1 individually.

(3) For $\mathbb{Z}_2$-vector spaces $E = E^0 \oplus E^1$ and $F = F^0 \oplus F^1$, the vector spaces $E \oplus F$, $E \otimes F$ and $\operatorname{Hom}(E, F)$ are regarded as $\mathbb{Z}_2$-vector spaces by $\epsilon_E \oplus \epsilon_F$, $\epsilon_E \otimes \epsilon_F$ and $f \mapsto \epsilon_F f \epsilon_E$, respectively. In other words, the $\mathbb{Z}_2$-graded structures are given by the direct sum decompositions

$$
\begin{aligned}
E \oplus F &= (E^0 \oplus F^0) \oplus (E^1 \oplus F^1), \\
E \otimes F &= (E^0 \otimes F^0 \oplus E^1 \otimes F^1) \oplus (E^0 \otimes F^1 \oplus E^1 \otimes F^0), \\
\operatorname{Hom}(E, F) &= (\operatorname{Hom}(E^0, F^0) \oplus \operatorname{Hom}(E^1, F^1)) \\
&\quad \oplus (\operatorname{Hom}(E^0, F^1) \oplus \operatorname{Hom}(E^1, F^0)).
\end{aligned}
$$

A representation of the Clifford algebra is defined to be a $\mathbb{Z}_2$-graded unitary representation in the following way.

DEFINITION 2.14. A $\mathbb{Z}_2$-graded unitary representation of the Clifford algebra $Cl(E)$ is a pair of a $\mathbb{Z}_2$-graded complex vector space R, equipped with a Hermitian metric, and an $\mathbb{R}$-algebra homomorphism $c_R : Cl(E) \to \mathrm{End}_{\mathbb{C}} R$ such that $c_R(v)$, for each v in E, is degree 1 and skew-Hermitian.

The condition that $\epsilon_R(v)$ is degree 1 is equivalent to that it anti-commutes with ϵ_R. Or the mapping $Cl(E) \times R \to R$ preserves $\mathbb{Z}_2$-degree. Namely, R is a representation preserving $\mathbb{Z}_2$-degree.

For a $\mathbb{Z}_2$-graded unitary representation, the degree 0 part and the degree 1 part have the same dimension.

A $\mathbb{Z}_2$-graded unitary representation is defined to be irreducible, if it does not contain, as a subrepresentation, a non-trivial $\mathbb{Z}_2$-graded unitary representation.

THEOREM 2.15.

(1) *If $n = 2m$, $Cl(E)$ has two irreducible $\mathbb{Z}_2$-graded unitary representations of dimension 2^m. These two representations are distinguished by the action of* vol, *which is one of $\pm(\sqrt{-1})^m \epsilon_R$. (In other words, there are two representations according to a choice of an orientation of E. They exchanged by shifting the degree.)*
(2) *If $n = 2m - 1$, $Cl(E)$ has a unique irreducible $\mathbb{Z}_2$-graded unitary representation of dimension 2^m.*

PROOF. $\mathbb{Z}_2$-graded representations of $Cl(E)$ are in one-to-one correspondence with unitary representations of $Cl(\mathbb{R}e_0 \oplus E)$. Hence the theorem reduces to Theorem 2.11. □

When the dimension of E is even, $Cl(E)$ has a unique irreducible unitary representation and two irreducible $\mathbb{Z}_2$-graded unitary representations. The relation among them is the following.

Let R be a (not necessarily irreducible) unitary representation of $Cl(E)$, for E of dimension $2m$. Choose and fix an orientation of E. Suppose that an orthonormal basis $\{e_1, e_2, \dots, e_{2m}\}$ respects the orientation. The element $\mathrm{vol} = e_1 e_2 \cdots e_{2m}$ is an element in $Cl(E)$ depending only on the orientation. Since $n = 2m$ is even, vol anti-commutes any elements in E. Noting that $\mathrm{vol}^2 = (-1)^m$, the only possibilities for eigenvalues of vol acting on R are $\pm(\sqrt{-1})^m$. Hence the Clifford action of $(\sqrt{-1}\,)^m\mathrm{vol}$ belongs to $H(R)$. We get the $\mathbb{Z}_2$-grading by eigenspace decomposition of the action of vol. There are two structures of $\mathbb{Z}_2$-graded unitary representations according to either eigenspace being regarded as degree 0.

In particular, from an irreducible unitary representation R, we get two irreducible $\mathbb{Z}_2$-graded unitary representations.

We summarize the discussion.

LEMMA 2.16. *Suppose that E is oriented and its dimension is even, $n = 2m$. For a unitary representation R of $Cl(E)$, we can define two different $\mathbb{Z}_2$-grading structures by $\epsilon_R = \pm(\sqrt{-1}\,)^m\mathrm{vol}$. For each grading, R becomes a $\mathbb{Z}_2$-graded unitary representation of $Cl(E)$. If R is irreducible, they are the only irreducible $\mathbb{Z}_2$-graded unitary representations of $Cl(E)$.*

If the dimension of E is odd, the action of vol on any $\mathbb{Z}_2$-graded unitary representation must be an isomorphism of degree 1, which commutes with the action of $Cl(E)$. Hence the $\mathbb{Z}_2$-graded unitary representation, obtained by exchanging the $\mathbb{Z}_2$-grading, is isomorphic to the original one.

In order to see whether a $\mathbb{Z}_2$-graded unitary representation R of $Cl(E)$ is the restriction of a representation of $Cl(\widetilde{E})$, for a strictly larger $\widetilde{E}$, we introduce $H_{\mathbb{Z}_2}(R)$ as follows:

$$H_{\mathbb{Z}_2}(R) = \{\gamma \in H(R) \mid \gamma\epsilon_R + \epsilon_R\gamma = 0\}.$$

Namely, $H_{\mathbb{Z}_2}(R)$ is $H(R)$ for a unitary representation R of $Cl(\mathbb{R}e_0 \oplus E)$.

3.1.2. *Examples of unitary representations.* We present some examples of $\mathbb{Z}_2$-graded unitary representations of the Clifford algebra.

First of all, we describe, explicitly, $\mathbb{Z}_2$-graded unitary representations of $Cl(E)$ in the case that the dimension of E is $1, 2, 4$. We also give a construction of unitary representations, using exterior algebras, for E carrying a complex structure.

EXAMPLE 2.17. Let $E = \mathbb{R}$, and $R = \mathbb{C}\oplus\mathbb{C}$, where $R^0 = R^1 = \mathbb{C}$. We define a real linear mapping $c_R : E \to \operatorname{End} R$ by

$$c_R(e_1) = \begin{pmatrix} 0 & -1 \\ 1 & 0 \end{pmatrix}.$$

Here $e_1 = 1$. Then R is an irreducible $\mathbb{Z}_2$-graded unitary representation of $Cl(\mathbb{R})$. It is not irreducible, if we forget the $\mathbb{Z}_2$-grading structure. In this example, we have

$$H_{\mathbb{Z}_2}(R) = \mathbb{R}\begin{pmatrix} 0 & 1 \\ 1 & 0 \end{pmatrix}.$$

Since $c_R(v)$ acts as a real matrix, R is a complexification of a $\mathbb{Z}_2$-graded real representation. In other words, the complex conjugation mapping $J : R \to R$ of $R = \mathbb{C} \oplus \mathbb{C}$ is a conjugate linear transformation that satisfies $J^2 = \mathrm{id}_R$ and commutes with the Clifford multiplication as well as the action of ϵ_R.

EXAMPLE 2.18. Let $E = \mathbb{R}^2$, and $R = \mathbb{C} \oplus \mathbb{C}$, where $R^0 = R^1 = \mathbb{C}$. We define a real linear mapping $c_R : E \to \operatorname{End} R$ by

$$c_R(e_1) = \begin{pmatrix} 0 & -1 \\ 1 & 0 \end{pmatrix}, \qquad c_R(e_2) = \begin{pmatrix} 0 & i \\ i & 0 \end{pmatrix}.$$

Here, e_1, e_2 is an orthonormal basis of $\mathbb{R}^2$. Then R is a irreducible $\mathbb{Z}_2$-graded unitary representation of $Cl(\mathbb{R}^2)$. (Another irreducible $\mathbb{Z}_2$-graded unitary representation is the one with exchanged grading.) It is still irreducible, after forgetting the $\mathbb{Z}_2$-grading structure. In this case, we have $H_{\mathbb{Z}_2}(R) = 0$. Forgetting the complex structure on R, we get a real representation $R_{\mathbb{R}}$ of $Cl(E)$. Denote it by $R_{\mathbb{R}}$. Then the set $H_{\mathbb{Z}_2}(R_{\mathbb{R}})$, which consists of all real symmetric transformations anti-commuting with $c_R(E)$ and ϵ_R, is not 0. A real linear transformation B on $\mathbb{C}$ is called a conjugate linear transformation, if it satisfies $Ba = \overline{a}B$, for any complex number a. A conjugate linear transformation on $\mathbb{C}$ is a symmetric transformation as real linear transformations. (A "reflection with respect to a line" on the complex plane $\mathbb{C}$ is an example.) It is easy to check that

$$H_{\mathbb{Z}_2}(R_{\mathbb{R}}) = \left\{ \begin{pmatrix} 0 & B \\ B & 0 \end{pmatrix} \middle| B \text{ is a conjugate linear transformation on } \mathbb{C} \right\}.$$

(See Lemma 4.19 in Chapter 4.)

In order to describe briefly the case that E is 4-dimensional, we review quarternion numbers. Let $\mathbb{H} = \mathbb{R} + \mathbb{R}i + \mathbb{R}j + \mathbb{R}k$ be the set of all quarternion numbers. This $\mathbb{H}$ is a non-commutative field and its relations are $i^2 = j^2 = k^2 = -1, \quad ij + jk = jk + kj = ki + ik = 0, \quad ij = k$. We consider it as a 2-dimensional vector space by the right multiplicative action of $\mathbb{C} = \mathbb{R} + \mathbb{R}i$. We take a basis $1, j$ and write $\mathbb{H} = \mathbb{C} + j\mathbb{C}$. Then the left multiplication by i, j, k is represented by the following matrices:

$$i = \begin{pmatrix} \sqrt{-1} & 0 \\ 0 & -\sqrt{-1} \end{pmatrix}, \quad j = \begin{pmatrix} 0 & -1 \\ 1 & 0 \end{pmatrix},$$
$$k = \begin{pmatrix} 0 & -\sqrt{-1} \\ -\sqrt{-1} & 0 \end{pmatrix}$$

These give a basis of the vector space of all skew-Hermitian matrices of trace 0, and, hence, $\mathbb{R}i + \mathbb{R}j + \mathbb{R}k$ is identified with the set of all skew-Hermitian matrices of trace 0.

EXAMPLE 2.19. Let $E = \mathbb{R}^4$ and $R = \mathbb{H} \oplus \mathbb{H}$, where $R^0 = R^1 = \mathbb{H}$. Denote by e_0, e_1, e_2, e_3 an orthonormal basis of $\mathbb{R}^4$ and define

$$\begin{aligned} c_R(e_0) &= \begin{pmatrix} 0 & -1 \\ 1 & 0 \end{pmatrix} \\ c_R(e_1) &= \begin{pmatrix} 0 & i \\ i & 0 \end{pmatrix} \\ c_R(e_2) &= \begin{pmatrix} 0 & j \\ j & 0 \end{pmatrix} \\ c_R(e_3) &= \begin{pmatrix} 0 & k \\ k & 0 \end{pmatrix}, \end{aligned}$$

which makes R an irreducible $\mathbb{Z}_2$-graded unitary representation. Note that the above action of quarternions is given by the multiplication of the matrix from the left, which commute with the right action by $\mathbb{H}$. Hence they are $\mathbb{H}$-linear. Forgetting the $\mathbb{Z}_2$-grading, it is still irreducible.

This R is not a complexification of any $\mathbb{Z}_2$-graded real representation. There is, however, a conjugate linear transformation, which commutes with the Clifford multiplication and ϵ_R and satisfies $J^2 = -\mathrm{id}_R$. Define J by the right multiplication by j. Then it indeed enjoys these properties. Existence of such a J is equivalent to the condition that R is a vector space over $\mathbb{H}$ and the representation of $Cl(E)$ is given by $\mathbb{H}$-linear transformations.

For an even dimensional Euclidean space, irreducible $\mathbb{Z}_2$-graded representations can be constructed using the exterior algebra.

We review the interior product operation as a preliminary.

DEFINITION 2.20. Let E be a real vector space equipped with a Euclidean metric. For an orthonormal basis $e_1, e_2, \ldots, e_n$, we define a Euclidean metric on $\Lambda^\bullet E$ such that the set of elements of the form $e_{k_1} \wedge e_{k_2} \wedge \cdots \wedge e_{k_a}$ is an orthonormal basis. For an element v of E, we define the interior product

$$v^{\lrcorner} : \wedge^k E \to \wedge^{k-1} E$$

by the transpose of the exterior product, of v, from the left

$$v^{\wedge} : \wedge^{k-1} E \to \wedge^k E.$$

Namely, $(v^{\wedge} u_0, u_1) = (u_0, v^{\lrcorner} u_1)$.

Intuitively, using the basis, $e_k^{\wedge}$ is the operation adding e_k on the left, and $e_k^{\lrcorner}$ is the one deleting e_k from the left. A fundamental property of the interior product is the following relation as linear operators on $\Lambda^\bullet E$:

$$v^{\wedge} v^{\lrcorner} + v^{\lrcorner} v^{\wedge} = |v|^2.$$

REMARK 2.21. Note that some articles adopt the opposite sign convention in the definition of the interior product.

The interior product is also defined, in a similar way, in the case of complex vector space with a Hermitian metric. But the "adjoint" is used instead of the "transpose". We use the Hermitian version of the interior product in the following example.

EXAMPLE 2.22. Suppose that E carries a structure of a complex vector space of complex dimension m. We denote by $E_{\mathbb{C}}$ the vector space E regarded as a complex vector space. Moreover, we assume that the Euclidean metric on E comes from a Hermitian metric on $E_{\mathbb{C}}$. We define a (complex) 2^m-dimensional $\mathbb{Z}_2$-graded unitary representation (R, c_R) by

$$R = \Lambda^\bullet E_{\mathbb{C}}, \quad c_R(v) = v^{\wedge} - v^{\lrcorner}.$$

The $\mathbb{Z}_2$-grading is defined by the parity of the degree in the exterior algebra. Since the dimension is 2^m, it is automatically irreducible as a $\mathbb{Z}_2$-graded unitary representation.

3.1.3. *Tensor product.* When the dimension of E is large, representations can be explicitly described by tensor products of representations of the Clifford algebra for lower dimensional vector spaces. We explain this fact.

In general, for two $\mathbb{Z}_2$-graded $\mathbb{R}$-algebras $A = A^0 \oplus A^1$ and $B = B^0 \oplus B^1$, the tensor product $A \otimes B$ becomes a $\mathbb{Z}_2$-graded $\mathbb{R}$-algebra as follows. The $\mathbb{Z}_2$-grading has been defined for vector spaces. The multiplicative structure is given by

$$(a \otimes b)(a' \otimes b') = (-1)^{(\deg b)(\deg a')} aa' \otimes bb',$$

where $a, a' \in A^0 \cup A^1$ and $b, b' \in B^0 \cup B^1$. Then we obviously have

LEMMA 2.23. $Cl(E_0 \oplus E_1) \cong Cl(E_0) \otimes Cl(E_1)$.

In general, for given $\mathbb{Z}_2$-graded representations R_A and R_B of A and B, respectively, we define a representation of $A \otimes B$ on $R_A \otimes R_B$ as follows:

$$(a \otimes b)(r_A \otimes r_B) = (-1)^{(\deg b)(\deg r_A)}(ar_A \otimes br_B),$$

where $r_A \in R_A^0 \cup R_A^1$, $r_B \in R_B^0 \cup R_B^1$, $a \in A^0 \cup A^1$ and $b \in B^0 \cup B^1$.

Irreducible representations of $Cl(E)$ can be constructed, using the following lemma, by the induction on the dimension of E.

LEMMA 2.24. *Suppose that at least one of E_0 and E_1 is even dimensional. For irreducible $\mathbb{Z}_2$-graded unitary representations R_{E_0} and R_{E_1} of $Cl(E_0)$ and $Cl(E_1)$, respectively, $R_{E_0} \otimes R_{E_1}$ is an irreducible $\mathbb{Z}_2$-graded unitary representation of $Cl(E_0) \otimes Cl(E_1)$, namely $Cl(E_0 \oplus E_1)$.*

PROOF. The irreducibility follows from the dimension of the vector space. □

EXAMPLE 2.25. When $E = \mathbb{R}^8$, we may regard E with the direct sum of two $\mathbb{R}^4$ and construct a $\mathbb{Z}_2$-graded irreducible representation of $Cl(E)$. Let R_0 and R_1 be irreducible representations of $Cl(\mathbb{R}^4)$. According to Lemma 2.16, they carry structures of $\mathbb{Z}_2$-grading. Then $R_0 \otimes R_1$ becomes an irreducible representation of $Cl(E) = Cl(\mathbb{R}^8)$ and is equipped with a $\mathbb{Z}_2$-grading by Lemma 2.16. An explicit construction in Example 2.19 implies that R_0 and R_1 are considered as vector spaces over quarternions. Namely, there are conjugate linear transformations J_0 and J_1 commuting with the Clifford multiplication and $\epsilon_{R_0}, \epsilon_{R_1}$, such that $J_0^2 = -1$ and $J_1^2 = -1$ hold. Then $J = J_0 \otimes J_1$ is a conjugate linear transformation commuting with the Clifford multiplication and $\epsilon_{R_0 \otimes R_1}$, such that $J^2 = 1$ holds. Namely, $R_0 \otimes R_1$ is considered as the complexification of a $\mathbb{Z}_2$-graded real representation of $Cl(E)$.

REMARK 2.26. In case that $\dim E = 0 \mod 8$, any irreducible representation of $Cl(E)$ is the complexification of some real representation and carries a $\mathbb{Z}_2$-grading compatible with the complexification.

This fact is evident from the construction by tensor product. In general, when a (complex) irreducible representation of a Clifford algebra is the complexification of some real representation, an element in the real representation space is sometimes called a **Majoranna spinor**. When a (complex) irreducible representation carries a $\mathbb{Z}_2$-grading, an element in either grading is sometimes called a **Weyl spinor**. In this terminology, for a Euclidean space with dimension divisible by 8, there exist Majoranna-Weyl spinors[1]. This fact is used in the proof of Bott-periodicity over $\mathbb{R}$ in Chapter 9 of the sequel.

Using the construction by tensor product and the induction on the dimension, we can easily see the following lemma.

LEMMA 2.27.

(1) *When the dimension of E is even, we denote by Δ_E a ($\mathbb{Z}_2$-graded) irreducible unitary representation. Then we have $Cl(E) \otimes \mathbb{C} \cong \mathrm{Hom}(\Delta_E, \Delta_E)$ (as $\mathbb{Z}_2$-graded objects).*
(2) *When the dimension of E is odd, we denote by $\Delta_E = \Delta_E^0 \oplus \Delta_E^1$ a $\mathbb{Z}_2$-graded irreducible unitary representation. Then we have $Cl(E) \otimes \mathbb{C} \cong \mathrm{Hom}(\Delta_E^0, \Delta_E)$.*

3.1.4. *Spin^c, Pin^c, Spin.* The orthogonal group $O(E)$ acts only on E, and does not act on representation spaces of $Cl(E)$ in a direct way. But we can find that there exists a certain central extension of $O(E)$, which acts on the irreducible representation space of $Cl(E)$ naturally, as follows.

Let Δ_E be a $\mathbb{Z}_2$-graded irreducible unitary representation of $Cl(E)$. (When the dimension of E is even, we fix one of two possibilities. Then this choice determines an orientation of E.) For an element O of the orthogonal group $O(E)$, O acts on $Cl(E)$ as an isomorphism of $\mathbb{Z}_2$-graded $\mathbb{R}$-algebra in a natural way. Hence, we can define a $\mathbb{Z}_2$-grading preserving representation of $Cl(E)$ on Δ_E by the composition of

$$Cl(E) \xrightarrow{O} Cl(E) \to \mathrm{End}\,\Delta_E.$$

When the dimension of E is odd, there is only one irreducible representation, which must be isomorphic to the original representation by a unitary transformation of Δ_E. When the dimension of E is even, if O respects the orientation of E (i.e., an element of $SO(E)$), then the

[1]In the case of the Minkowski space, the corresponding condition is that $\dim E \equiv 2 \bmod 8$. It differs from the Euclidean case by 2. This fact is of basic importance in physics.

obtained representation is isomorphic to the original representation. But, the isomorphism is not unique. However, by Schur's lemma, it is unique up to the multiplication (by complex numbers of absolute value 1). Consider all unitary transformations of Δ_E, which are obtained by some O as above. Then we find the following

LEMMA 2.28. *When the dimension of E is odd, there is the following central extension*

$$0 \to U(1) \to Pin^c(E) \to O(E) \to 0,$$

and $Pin^c(E)$ acts on Δ_E. When the dimension of E is even, there is the following central extension

$$0 \to U(1) \to Spin^c(E) \to SO(E) \to 0,$$

and $Spin^c(E)$ acts on Δ_E.

REMARK 2.29. As in the above lemma, when the action is defined after taking a certain central extension, we call such an action a projective action. Intuitively, the general principle is the following: "Given an irreducible representation, which is rigid under deformations, of an algebra, then the automorphism group of the algebra acts on the irreducible representation in a projective way."

When the dimension of E is even, we obtain the same $Spin^c(E)$, whichever of the two irreducible representations we choose.

When the dimension of E is odd, we denote by $Spin^c(E)$ the central extension constructed using only O in $SO(E)$.

We shall give an explicit description of $Spin^c(E)$ constructed above. We can show the following statement by, for example, the induction of the dimension of E.

LEMMA 2.30. *Denote by $Spin(E)$ the double covering of $SO(E)$. Then we have*

$$Spin^c(E) \cong Spin(E) \times_{\{\pm 1\}} U(1).$$

In particular, $Spin(E)$ acts on Δ_E.

The fundamental group of $SO(E)$ is known to be isomorphic to $\mathbb{Z}$, when the dimension of E is 2 and $\mathbb{Z}_2$, when the dimension of E is greater than or equal to 3. Since double coverings are classified by surjective homomorphisms from the fundamental group to $\mathbb{Z}_2$, there exists a unique double covering space of $SO(E)$.

Since $Spin(E)$ and $SO(E)$ have the same Lie algebra $\mathfrak{spin}(E) \cong \mathfrak{so}(E)$, $\mathfrak{so}(E)$ acts on Δ. More generally, the following lemma tells us that $\mathfrak{so}(E)$ acts on any unitary representation of $Cl(E)$ in a natural way. Any representation is certainly a direct sum of irreducible representations, thus this is obvious from Lemma 2.30. But the following lemma gives an explicit description of the action.

LEMMA 2.31. *Identify $\mathfrak{so}(E)$ and $\bigwedge^2 E$ to define the linear mapping ϕ from $\mathfrak{so}(E)$ to $Cl(E)$ by assigning $uv - vu$ to $u \wedge v$. Define the Lie bracket on $Cl(E)$ by the commutator $[a, b] = ab - ba$. Then ϕ preserves the Lie bracket.*

In a precise sense, the statement does not make sense unless we fix the identification of $\mathfrak{so}(E)$ and $\bigwedge^2 E$. We omit the detail.

DEFINITION 2.32. Let E be an oriented vector space equipped with a Euclidean metric. We define a spinor representation of $Spin(E)$ or $Spin^c(E)$ by $\mathbb{Z}_2$-graded irreducible unitary representation Δ_E of $Cl(E)$. Here, when the dimension of E is even $n = 2m$, we assume that the action of $e_1 e_2 \cdots e_n$ coincides with the action of $(\sqrt{-1}\,)^m \epsilon_{\Delta_E}$, where $(e_1, e_2, \ldots, e_n)$ is an oriented orthonormal basis of E. When the dimension of E is odd, a larger group $Pin^c(E)$ acts on Δ_E.

When we identify $E = \mathbb{R}^n$ using coordinates, we write $Spin(n)$, $Spin^c(n)$ for $Spin(E)$, $Spin^c(E)$, respectively.

3.2. $\mathbb{Z}_2$-graded Clifford modules. We shall extend the construction on linear algebra in the previous section to families parametrized by a manifold X. For an n-dimensional real vector bundle T with a Euclidean metric on X, we consider the Clifford algebra $Cl((T)_x)$ on $(T)_x$ at each point x of X. Set $Cl(T)$ the vector bundle obtained by the family $Cl((T)_x)$ for all x.

DEFINITION 2.33. For a real vector bundle T with a Euclidean metric on X, (W, c_W) is called a **Clifford module over (X, T)**, if W is a complex vector bundle with a Hermitian metric and c_W is a homomorphism from T to $\mathrm{End}_{\mathbb{C}}\, E$ as real vector bundles so that the following conditions are satisfied:

(1) c_W extends to an algebra homomorphism $(Cl(T))_x \to (\mathrm{End}_{\mathbb{C}}\, E)_x$ fiberwisely.
(2) The image of c_W is contained in the space of skew Hermitian endomorphisms of E.

We call c_W the **Clifford multiplication** on W.

When T satisfies appropriate conditions, we can classify Clifford modules over (X, T).

DEFINITION 2.34.

(1) We say that T has a spin structure, if the structure group of T can be lifted to $Spin(n)$. Such a lift is called a **spin structure** of T.
(2) We say that T has $Spin^c$-structure, if the structure group of T can be lifted to $Spin^c(n)$. Such a lift is called a ***Spin^c*-structure** of T.
(3) When the rank of T is odd, we say that T has a Pin^c-structure, if the structure group of T can be lifted to $Pin^c(n)$. Such a lift is called a ***Pin^c*-structure** of T.

Here is an explanation. Consider the homomorphism

$$Spin(n) \to Spin^c(n) \to SO(n) \to O(n)$$

between compact Lie groups. Since T is equipped with a Euclidean metric, there is a principal $O(n)$-bundle P such that $T = P_0 \times_{O(n)} \mathbb{R}^n$. For example, a spin structure is a pair of a principal $Spin(n)$-bundle P_1 and an isomorphism $T \cong P_1 \times_{Spin(n)} \mathbb{R}^n$. It is equivalent to a pair of P_1 and an isomorphism $P_0 \cong P_1 \times_{Spin(n)} O(n)$.

In order to carry a spin structure, resp. $Spin^c$-structure, T must be orientable. Namely, it is necessary that the structure group is reduced to $SO(n)$ from $O(n)$. If a spin structure exists, a $Spin^c$-structure is naturally obtained by the homomorphism $Spin(n) \to Spin^c(n)$.

LEMMA 2.35.

(1) *When the rank of T (over $\mathbb{R}$) is an even number $2m$, there is a one-to-one correspondence between $Spin^c$-structures of T and isomorphism classes of Z_2-graded Clifford modules over (X, T) of rank 2^m (over $\mathbb{C}$). (In particular, if a $\mathbb{Z}_2$-graded Clifford module of rank 2^m is given, we can give an orientation of T.)*
(2) *If the rank of T is an odd number $2m - 1$, there is a one-to-one correspondence between Pin^c-structures of T and isomorphism classes of $\mathbb{Z}_2$-graded Clifford modules over (X, T) of rank 2^m.*

PROOF. We give a proof in the case of even dimension. The other case is similar. For a given $Spin^c$-structure, the pair of the principal

$Spin^c(2m)$-bundle and its associated vector bundle with fiber $\Delta_{\mathbb{R}^{2m}}$ is a $\mathbb{Z}_2$-graded Clifford module. Conversely, suppose that there exists a $\mathbb{Z}_2$-graded Clifford module W of rank 2^m. Pick an isomorphism O from $\mathbb{R}^{2m}$ to $(T)_x$ preserving the Euclidean metric. Then $Cl(\mathbb{R}^{2m})$ is isomorphic to $Cl((T)_x)$. Using this isomorphism, $Cl(\mathbb{R}^{2m})$ acts on $(W)_x$. Since this representation is irreducible, Schur's lemma implies that there is an isomorphism from $\Delta_{\mathbb{R}^{2m}}$ to $(W)_x$ as unitary representations of $Cl(\mathbb{R}^{2m})$. When it preserves the $\mathbb{Z}_2$-grading, we interpret "O is orientation preserving" and when it exchanges the $\mathbb{Z}_2$-grading, we interpret "O is orientation reversing". In this way, we can give an orientation on T. (One of these should occur, since there are only two ways of giving a $\mathbb{Z}_2$-grading.) For O, which is orientation preserving, an isomorphism from $\Delta_{\mathbb{R}^{2m}}$ to $(W)_x$ is determined up to the freedom of $U(1)$. Denote by $(P)_x$ the totality of all isomorphisms constructed from orientation preserving O as above. Then $Spin^c(2m)$ acts on $(P)_x$ on the right transitively and freely. The bundle P, which is obtained from the family of $(P)_x$ parametrized by x, is the desired principal $Spin^c(2m)$-bundle on X. □

DEFINITION 2.36. For a $Spin^c$-structure, the vector bundle with fiber $\Delta_{\mathbb{R}^{2m}}$ associated with the principal $Spin^c$-bundle is called the **spinor bundle**. For a Pin^c-structure of odd dimension, we define the spinor bundle in a similar way.

EXAMPLE 2.37. When T carries a structure of a complex vector bundle, we can make $\bigwedge^\bullet T_{\mathbb{C}}$ a $\mathbb{Z}_2$-graded Clifford module (cf. Example 2.22). This is one of $Spin^c$-structure of T.

PROPOSITION 2.38.

(1) *Suppose that the rank of T is even and T carries a $Spin^c$-structure. Denote by S_T the corresponding spinor bundle. There is a one-to-one correspondence between isomorphism classes of $\mathbb{Z}_2$-graded Clifford modules $W = W^0 \oplus W^1$ over (X, T) and isomorphism classes of $\mathbb{Z}_2$-graded vector bundle $F = F^0 \oplus F^1$ on X by $W = F \otimes S_T$.*

(2) *Suppose that the rank of T is odd and T carries a Pin^c-structure. Denote by S_T the corresponding spinor bundle. There is a one-to-one correspondence between isomorphism classes of $\mathbb{Z}_2$-graded Clifford modules $W = W^0 \oplus W^1$ over (X, T) and isomorphism classes of vector bundles F^0 over X by $W = F^0 \otimes S_T$.*

In the case that the rank is odd, we can also regard F^0 as a $\mathbb{Z}_2$-graded vector bundle such that the part of degree 1 is 0.

PROOF. When X consists of a point, the statement clearly follows from classification of irreducible $\mathbb{Z}_2$-graded representations. (Similar to unitary representations of finite groups, unitary representations of Clifford algebras are completely reducible.) This proposition is its direct extension to vector bundles.

Set $F^0 = \mathrm{Hom}^0_{Cl(T)}(S_T, W)$. Here denote by $\mathrm{Hom}^0_{Cl(T)}$ the totality of homomorphisms, which commutes with $Cl(T)$-action and preserves the $\mathbb{Z}_2$-grading. Similarly, denote by $F^1 = \mathrm{Hom}^1_{Cl(T)}(S_T, W)$ the totality of homomorphisms, which exchages the $\mathbb{Z}_2$-grading. When the rank is even, $F = F^0 \oplus F^1$ is the desired object. When the rank is odd, F^0 is the desired object. □

We shall give another expression of $\mathbb{Z}_2$-graded Clifford modules in terms of principal bundles.

Firstly, we consider the case that T is oriented and the rank is an even number $2m$.

$U(r_0) \times U(r_1) \times Spin^c(2m)$ acts on $(\mathbb{C}^{r_0} \oplus \mathbb{C}^{r_1}) \otimes \Delta_{\mathbb{R}^{2m}}$ in a natural way. The action of the subgroup $\{(z, z, z^{-1}) \mid z \in U(1)\}$ is trivial. Denote this subgroup by $U(1)$ and consider the homomorphism

$$\frac{U(r_0) \times U(r_1) \times Spin^c(2m)}{U(1)} \to SO(2m)$$

taking the third factor. If the structure group of T is lifted to the group on the left hand side, we can construct the associated vector bundle with fiber $(\mathbb{C}^{r_0} \oplus \mathbb{C}^{r_1}) \otimes \Delta_{\mathbb{R}^{2m}}$. This bundle becomes a $\mathbb{Z}_2$-graded Clifford module in a natural way. Conversely, by a similar argument to the proof of Proposition 2.38, any $\mathbb{Z}_2$-graded Clifford module can be obtained in this way. A similar argument works in the case that the rank is odd. Let us summarize it.

PROPOSITION 2.39.

(1) *Suppose that the rank of T is $2m$ and T is oriented. $\mathbb{Z}_2$-graded Clifford modules over (X, T) are in one-to-one correspondence with liftings of the structure group of T with respect to the homomorphism*

$$\frac{U(r_0) \times U(r_1) \times Spin^c(2m)}{U(1)} \cong \frac{U(r_0) \times U(r_1) \times Spin(2m)}{\{\pm 1\}} \to SO(2m).$$

(2) *Suppose that the rank of T is $2m-1$. $\mathbb{Z}_2$-graded Clifford modules over (X,T) are in one-to-one correspondence with liftings of the structure group of T with respect to the homomorphism*

$$\frac{U(r)\times Pin^c(2m-1)}{U(1)} \to O(2m-1).$$

We omit the case that the rank of T is even but T is not orientable, since the situation is slightly complicated.

EXERCISE 2. When the dimension of E is $2m$, we denote by $\tilde{\Delta}_E$ the direct sum of two irreducible $\mathbb{Z}_2$-graded unitary representations of $Cl(E)$, which are not isomorphic to each other. Using the action of $O(E)$, construct the extension

$$1 \to U(1)\times U(1) \to Pin^{cc}(E) \to O(E) \to 1$$

and the action of $Pin^{cc}(E)$ on $\tilde{\Delta}_E$ (not a central extension). Using this, study the case that the rank of T is even but T has no orientation.

EXERCISE 3.

(1) Denote by $\tilde{U}(m)$ the totality of elements (U,z) of $U(m)\times U(1)$ such that $\det U = z^2$. Show that $\tilde{U}(m)$ is isomorphic to the fiber product of the natural homomorphisms $Spin(2m)\to SO(2m)$ and $U(m)\to SO(2m)$.

(2) Suppose that T carries a structure of a complex vector bundle of complex rank m and set $K=\bigwedge^m T^*_{\mathbb{C}}$. Show that spin structures of T are in one-to-one correspondence with complex line bundles L such that $L^{\otimes 2}=K$. Show also that the spinor bundle associated to the spin structure corresponding to L is isomorphic to $L\otimes\bigwedge^\bullet T_{\mathbb{C}}$.

3.3. Operators of Dirac Type. We shall define the operator of Dirac type, when T in the previous section is isomorphic to T^*X.

Firstly, we define differential operators and their principal symbols on manifolds, in general.

DEFINITION 2.40. Let W_0, W_1 be complex vector bundles on a manifold X. (Our argument also works for real vector bundles, but we adopt the above setting for fixing notation.) A linear mapping $D:\Gamma(W_0)\to\Gamma(W_1)$ is called a **linear differential operator of order at most k**, if the following conditions are satisfied.

(1) For any element s of $\Gamma(W_0)$, the support of Ds is contained in the support of s.

(2) Let U be an open subset of X, $x = (x^1, x^2, \dots, x^n)$ the coordinates on U and $\phi_0 : W_0 \to \mathbb{C}^{r_0}$, $\phi_1 : W_1 \to \mathbb{C}^{r_1}$ local trivializations of W_0, W_1. For a multi-index $\alpha = (\alpha_1, \alpha_2, \dots, \alpha_n)$, there exists a smooth function $a_\alpha(x)$ on U with values in $\mathrm{Hom}(\mathbb{C}^{r_0}, \mathbb{C}^{r_1})$ such that

$$(Ds)(x) = \sum_{|\alpha| \le k} a_\alpha(x)(\partial^\alpha s)(x) \tag{2.1}$$

holds for any section $s(x)$ of W_0 with support in U. Here we write

$$|\alpha| = \sum_i \alpha_i \quad \text{and} \quad \partial^\alpha = \prod_i \partial_{x^i}^{\alpha}$$

In the above definition, local expression of the differential operator is transformed in a complicated way under coordinate changes and changes of local trivializations.[2] However, it is easy to see that the part of the highest order k enjoys a simple transformation. We only state the conclusion. For each x, we can define the mapping

$$\sigma_k(D)_x : (T^*X)_x \to \mathrm{Hom}(W_0, W_1)_x$$

by

$$\sum_i u_i dx^i \mapsto \sum_{|\alpha|=k} a_\alpha(x) \prod u_i^{\alpha_i}$$

using the expression in coordinates and local trivializations. This mapping changes smoothly in x. Set

$$\sigma_k(D) : T^*X \to \mathrm{Hom}(W_0, W_1)$$

the family obtained by changing x. $\sigma_k(D)$ is a homogeneous polynomial of order k with values in the complex vector bundle $\mathrm{Hom}(W_0, W_1)$ over the real vector bundle T^*X. It is linear (over $\mathbb{R}$) only when $k = 1$. In general, it satisfies the homogeneity of degree k for the real number t, i.e., $\sigma_k(D)(tu) = t^k \sigma_k(D)(u)$.

DEFINITION 2.41. For a linear differential operator D of order k, we call $\sigma_k(D)$ the principal symbol of **principal symbol** D.

REMARK 2.42. In some references, $\sigma_k(D)$ multiplied by $(\sqrt{-1}\,)^k$ is called the principal symbol. This is equivalent to replacing ∂_{x_i} in the local expression of D by $\partial_{x_i}/\sqrt{-1}$. For the definition of Hermiteness for the Dirac operator, this definition may be suitable, since it is

[2] It is the coordinate change of the "jet bundle".

compatible with the Hermiteness of the principal symbol. But we defined the principal symbol without the multiplication by $\sqrt{-1}$ in this book, since we will also discuss principal symbols of linear differential operators on real vector bundles.

From this definition, we clearly find the following additivity and composition law.

LEMMA 2.43. *Let $D_0, D_0' : \Gamma(W_0) \to \Gamma(W_1)$ be both linear differential operators of order at most k_0 and $D_1 : \Gamma(W_1) \to \Gamma(W_1)$ a linear differential operator of order at most k_1. Then the sum $D_0 + D_0'$ is a linear differential operator of order at most k_0 and the composition $D_1 D_0$ is a linear differential operator of order at most $k_0 + k_1$ such that*

$$\sigma_{k_0}(D_0+D_0')=\sigma_{k_0}(D_0)+\sigma_{k_0}(D_0'), \quad \sigma_{k_0+k_1}(D_1 D_0)=\sigma_{k_1}(D_1)\sigma_{k_0}(D_0)$$

holds.

LEMMA 2.44. *Let σ be a homogeneous polynomial of degree k on T^*X with values in $\mathrm{Hom}(W_0, W_1)$. Then there exists a linear differential operator of order at most k such that its principal symbol is σ.*

PROOF. We can show existence on a sufficiently small neighborhood U_x at each point x using coordinates and local trivializations. In general, we pick a locally finite refinement $\{U_j\}$ of $\{U_x\}$. Let D_j be a linear differential operator of order at most k on U_j. Using a partition of unity ρ_j^2, we take the average $D = \sum \rho_j D_j \rho_j$. Regard the multiplication by ρ_j as a differential operator of order 0. Then its principal symbol is ρ_j itself. Hence, by the additivity and the composition law of the principal symbol, we find that

$$\sigma_k(D) = \sum \rho_j \sigma_k(D_j)\rho_j = \sum \rho_j^2 \sigma = \sigma.$$

□

When X is a Riemannian manifold, we call a Clifford module over (X, T^*X) simply a **Clifford module on X**.

DEFINITION 2.45. Let X be a Riemannian manifold. When (W, c_W) is a $\mathbb{Z}_2$-graded Clifford module on X, we call a linear differential operator

$$D : \Gamma(W) \to \Gamma(W)$$

of order at most 1 an **operator of Dirac type on W** if the following conditions are satisfied:

- D shifts the grading of $\Gamma(W)$. Namely, D is of degree 1.
- $\sigma_1(D) = c_W$.
- D is formally self-adjoint.

We explain the last condition. In general, we assume that a manifold X of dimension n is equipped with a volume element $d\,\mathrm{vol}$. Namely, $d\,\mathrm{vol}$ is an element of $\Omega^n(X, \theta_X)$, which is "positive" everywhere. Let W_0, W_1 be complex vector bundles on X with Hermitian metrics. The formal adjoint of the linear differential operator $D : \Gamma(W_0) \to \Gamma(W_1)$ is defined to be the linear differential operator $D^* : \Gamma(W_1) \to \Gamma(W_0)$ such that

$$\int_X d\,\mathrm{vol}\langle Ds_0, s_1\rangle_{W_1} = \int_X d\,\mathrm{vol}\langle s_0, D^* s_1\rangle_{W_0}$$

holds for compactly supported sections s_0, s_1 of W_0, W_1, respectively. It is clear from this definition that it is uniquely determined, if it exists.

Existence easily follows from local expressions. Namely, let (2.1) be a local expression of D and write $d\,\mathrm{vol} = g(x)^{1/2}|dx_1 \cdots dx_n|$, $g(x) > 0$. Then the formal adjoint of D is given by

$$D^* s_1 = g^{-1/2} \sum_{|\alpha| \le k} (-1)^{|\alpha|}\partial^\alpha (g^{1/2} a_\alpha^* s_1).$$

Here, a_α^* is the adjoint matrix of a_α.

If X is equipped with a Riemannian metric, it determines the volume element $d\,\mathrm{vol}$ in a canonical way. We use it to define the formal adjoint.

D is said to be **formally self-adjoint**, if $W_0 = W_1$ and $D = D^*$.

LEMMA 2.46. *For any Clifford module over (X, T^*X), there exists an operator of Dirac type on it.*

PROOF. By Lemma 2.44, there exists D, which satisfies all conditions for operators of Dirac type except the condition on formal self-adjointness. Since the Clifford multiplication c_W is skew Hermitian, the formal adjoint D^* of D is the operator of order 1, which has the same principal symbol as D. Thus $(D + D^*)/2$ becomes formally self-adjoint and is an operator of Dirac type. □

REMARK 2.47. In some references, "the Dirac operator " is defined to be the operator, which satisfies a little stronger condition than Definition 2.45. In this book, we use the terminology "operator of Dirac type" to avoid confusion.

Before closing this section, we note some delicate points on spin structures and $Spin^c$-structures.

Let T be a real vector bundle on X. We assume that T is oriented, but is not equipped with a Euclidean metric. In such a situation, can we define the notion of spin structures or $Spin^c$-structures on T without introducing a Euclidean metric? Once we fix a metric on T, it is possible for each choice. Since any two metrics are related under an automorphism of T and such automorphisms are homotopic to the identity, it is clearly possible to define these notions, if we allow the freedom of homotopy. However, since each Dirac operator is defined on each Clifford module, we would like to exclude the freedom of homotopy, if possible. As a conclusion, we can define spin structures or $Spin^c$-structures on the level of principal bundles without using a Euclidean metric, but the definition of Clifford modules requires a Euclidean metric. In fact, $GL_n^+(\mathbb{R})$ (+ stands for orientation preserving) has the same homotopy type as $SO(n)$. Denote by $\widetilde{GL_n^+}(\mathbb{R})$ its double covering. We can define a spin structure of T by the fact that the structure group is lifted to $\widetilde{GL_n^+}(\mathbb{R})$. We can define $Spin^c$-structures of T in a similar way using $\widetilde{GL_n^+}(\mathbb{R}) \times_{\pm 1} U(1)$. Then we do not need to use a Euclidean metric of T. However (except in the case that $n = 1$), the action of $Spin^c(n)$ on irreducible unitary representations of $Cl(\mathbb{R}^n)$ does not extend to the action of these non-compact Lie groups. Therefore we need to fix a Euclidean metric in order to define spinor bundles as associated vector bundles. There is no canonical isomorphism between two associated spinor bundles corresponding to distinct two Euclidean metrics. It is only possible to get a family of infinitely many isomorphisms, which are homotopic to one another.

For example, when we consider the question of how solutions of the operator of Dirac type change when we vary Riemannian metrics on X, it is not possible to think naively that the function space, where the solutions belong to, is determined before fixing a Riemannian metric.

4. Elliptic Differential Operators Appearing in Geometry and Operators of Dirac Type

Certain quantities with geometric significance may be interpreted as the dimension of the solutions of some operators of Dirac type.

4.1. de Rham Complex. Let X be an n-dimensional manifold. In this section, we shall study the **de Rham complex** $(\Omega^\bullet(X), d)$. The definition of this complex only requires topological information. Hence the cohomology of this complex

$$H^k(X) = \frac{\operatorname{Ker}(d : \Omega^k(X) \to \Omega^{k+1}(X))}{\operatorname{Im}(d : \Omega^{k-1}(X) \to \Omega^k(X))}$$

is also a topological object. This is the **de Rham cohomology** of X.

Now we pick an arbitrary Riemannian metric on X, then the cohomology is expressed as the kernel of a certain operator of Dirac type. In the previous section, we defined the Clifford module as a structure on a complex vector bundle. It may be clear that we can extend the definition to real vector bundles in a similar way. We call it a **real Clifford module**. After tensoring with $\mathbb{C}$, we obtain a Clifford module in the sense of the previous section. However, there are cases where the argument over $\mathbb{R}$ is more natural. The de Rham complex is a typical example of this kind.

The principal symbol $\sigma_1(d) : T^*X \to \operatorname{End} \bigwedge^\bullet T^*X$ of the differential operator d of the first order coincides with the wedge product by $v \mapsto v^\wedge$ from the left. Using the Riemannian metric, $\bigwedge^\bullet T^*X$ is equipped with a Euclidean metric.

Then we can define the formal adjoint d^*. The principal symbol $\sigma_1(d^*)$ of d^* is $v \mapsto -v^\lrcorner$. The minus sign appears here, since the formal adjoint of the first order differentiation $\partial/\partial x$ is $-\partial/\partial x$.

Define the first order linear differential operator D on $W = \bigwedge^\bullet T^*X$ by $d + d^*$. Then it is a formally self-adjoint operator exchanging the parity of degree and satisfies

$$\sigma_1(D)(v) = v^\wedge - v^\lrcorner.$$

We introduce the $\mathbb{Z}_2$-grading on $\bigwedge^\bullet T^*X$ by the parity of the degree of the exterior product.

Then it becomes a $\mathbb{Z}_2$-graded real Clifford module over (X, T^*X) by $c(v) = v^\wedge - v^\lrcorner$. Thus we regard D as an operator of Dirac type.

A significant property of the exterior differentiation is $d^2 = 0$. The following properties follow from this fact.

- $(d^*)^2 = 0$.
- $D^2 = dd^* + d^*d$ preserves the degree of differential forms.
- D^2 commutes with d and d preserves each eigenspace of D^2.

If a section s of $W = \bigwedge^\bullet T^*X$ satisfies $D^2 s = 0$, we have

$$\begin{aligned} 0 &= \int_X d\,\mathrm{vol}\langle D^2 s, s\rangle_W \\ &= \int_X d\,\mathrm{vol}(\langle dd^* s, s\rangle_W + \langle d^* ds, s\rangle_W) \\ &= \int_X d\,\mathrm{vol}(|d^* s|_W^2 + |ds|_W^2), \end{aligned}$$

hence $Ds = ds = d^* s = 0$. Conversely, it is clear that $Ds = D^2 s = 0$ if $ds = d^* s = 0$. Therefore we obtain a mapping from $\operatorname{Ker} D^2$ to $H^\bullet(X)$. We prove the following theorem in Chapter 5, §7.

THEOREM 2.48. *$\operatorname{Ker} D^2 \to H^\bullet(X)$ is an isomorphism. This space is finite dimensional.*

$\operatorname{Ker} D^2$ depends on the Riemannian metric, but its dimension does not depend on the choice of the Riemannian metric.

For a local system θ on X, a similar argument works for the **de Rham complex with coefficients in θ** $(\Omega^\bullet(X,\theta), d)$. The cohomology of this complex

$$H^k(X,\theta) = \frac{\operatorname{Ker}(d : \Omega^k(X,\theta) \to \Omega^{k+1}(X,\theta))}{\operatorname{Im}(d : \Omega^{k-1}(X,\theta) \to \Omega^k(X,\theta))}$$

is called the **de Rham cohomology with coefficients in θ**. Once we pick a Riemannian metric on the closed manifold X, then we can express it as Ker of the operator $D = d + d^*$ of Dirac type.

A particularly important case is that θ is the local system θ_X of orientations of X. The de Rham complex with coefficients in θ_X and its associated operator of Dirac type is closely related to what will be defined in Chapter 3, §5. In the proof of the index theorem given in this book, the supersymmetric harmonic oscillator is one of the key ingredients.

4.2. Dolbeault Complex. Suppose that a closed manifold X carries a structure of a complex manifold of complex dimension m. When we regard X as a complex manifold, we denote it by $X_{\mathbb{C}}$. TX carries a structure of a complex vector bundle. Denote by J the multiplicative action by $\sqrt{-1}$ on it. Extend the action J to $TX \otimes \mathbb{C}$ in $\mathbb{C}$-linear fashion. Decompose it into eigenspaces of J. Then the part, on which $J = \sqrt{-1}$, is isomorphic to $T_{\mathbb{C}}X$ as complex vector spaces (by the projection to TX taking the real component), and the part, on which $J = -\sqrt{-1}$, is isomorphic to $\overline{T_{\mathbb{C}}X}$. From this, we obtain the decomposition

$$\bigwedge^{\bullet} T^*X \otimes \mathbb{C} = \bigoplus_{p,q} \bigwedge^{p,q} T^*X_{\mathbb{C}}, \quad \bigwedge^{p,q} T^*X_{\mathbb{C}} \cong (\bigwedge^{p} T^*_{\mathbb{C}}X) \otimes (\bigwedge^{q} \overline{T^*_{\mathbb{C}}X}).$$

For sections of the bundle, we write

$$\Omega^{\bullet}(X) \otimes \mathbb{C} = \bigoplus_{p,q} \Omega^{p,q}(X_{\mathbb{C}})$$

for the corresponding decomposition.

The exterior differentiation $d \otimes \mathbb{C}$ is also decomposed according to the decomposition on the right hand side. In particular, we denote by $\overline{\partial}$ the component, which maps $\bigwedge^{p,q} T^*X_{\mathbb{C}}$ to $\bigwedge^{p,q+1} T^*X_{\mathbb{C}}$.

In the above construction, we used the structure of a complex vector bundle on TX, but we did not use the fact that it comes from the complex structure of X. However, using the fact that X is a complex manifold, we find that $\overline{\partial}^2 = 0$. From this, we can define the **Dolbeault complex** $(\Omega^{0,\bullet}(X_{\mathbb{C}}), \overline{\partial})$.

The Dolbeault complex is a counterpart of the de Rham complex in the realm of complex manifolds. As the de Rham complex is defined with coefficients in local systems, the Dolbeault complex is defined with coefficients in holomorphic vector bundles. For a holomorphic vector bundle F on X, set

$$\Omega^{0,\bullet}(X_{\mathbb{C}}, F) = \bigoplus_{q=0}^{m} \Gamma(\bigwedge^{0,q} T^*X_{\mathbb{C}} \otimes F) \Gamma(\bigwedge^{\bullet} \overline{T^*X_{\mathbb{C}}} \otimes F).$$

Using local trivializations of F such that transition functions are holomorphic, the **Dolbeault complex with coefficients in F** $(\Omega^{0,\bullet}(X_{\mathbb{C}}, F), \overline{\partial}_F)$ is defined in a natural way. Its cohomology

$$H^{0,q}(X_{\mathbb{C}}, F) = \frac{\operatorname{Ker}(\overline{\partial}_F : \Omega^{0,q}(X_{\mathbb{C}}, F) \to \Omega^{0,q+1}(X_{\mathbb{C}}, F))}{\operatorname{Im}(\overline{\partial}_F : \Omega^{0,q-1}(X_{\mathbb{C}}, F) \to \Omega^{0,q}(X_{\mathbb{C}}, F))}$$

depends only on the complex structure of X and the holomorphic structure of F.

In a parallel way to the case of the de Rham complex, we can express this cohomology in relation to a certain operator of Dirac type, if we pick an arbitrary Hermitian metric on $T_{\mathbb{C}}X$, F.

We equip the complex vector bundle $W = \bigwedge^{\bullet} \overline{T^*_{\mathbb{C}}X} \otimes F$ with the structure of a Clifford module over (X, T^*X) by $c(v) = v^{\wedge} - v^{\lrcorner}$. (Note that $\overline{T^*_{\mathbb{C}}X}$ is the same as T^*X as geometric objects.) Define the $\mathbb{Z}_2$-grading by the parity of degree of the exterior product, it becomes a $\mathbb{Z}_2$-graded Clifford module.

Set $D = \overline{\partial}_F + (\overline{\partial}_F)^*$, then D is an operator of Dirac type on W.

Parallel to the case of the de Rham complex, we will prove the following theorem in Chapter 5, §7.

THEOREM 2.49. *The natural mapping* $\operatorname{Ker} D^2 \to H^{0,q}(X_{\mathbb{C}}, F)$ *is an isomorphism. This space is finite dimensional.*

4.3. Signature. Suppose that X is an oriented closed manifold of even dimension and set $n = 2m$ the dimension of X. Then the bilinear form

$$H^m(X) \times H^m(X) \to \mathbb{R},$$

which is defined by the exterior product and integration, is symmetric, if m is even and skew symmetric, if m is odd.

DEFINITION 2.50. When m is even, the **signature** $\operatorname{sign} X$ of the $2m$-dimensional oriented closed manifold X is defined to be the difference between the dimension of a maximal positive definite subspace of $H^m(X)$ and the dimension of a maximal negative definite subspace.

When m is even, the signature of X is a topological invariant of X.

It is suitable for theoretical discussion to extend the symmetric or skew symmetric bilinear form to the total de Rham cohomology not only its degree m-part for theoretical discussion. Let u be a k-differential form and v a $(n-k)$-differential form. Consider the correspondence

$$(u, v) \mapsto (-1)^{\frac{1}{2}\{k(k+1)-m(m+1)\}} \int_X u \wedge v.$$

It induces a bilinear form

$$Q_{\text{sign}} : H^\bullet(X) \times H^\bullet(X) \to \mathbb{R},$$

which coincides with the previous one in the degree m, is symmetric, if $m = n/2$ is even, and skew symmetric, if it is odd. We can easily see this fact using

$$\frac{k(k+1)}{2} + \frac{l(l+1)}{2} \equiv kl + m, \quad \bmod 2,$$

when $k + l = 2m$.

LEMMA 2.51. *When m is even, the signature* $\operatorname{sign} X$ *of* X *coincides with the signature of the symmetric bilinear form* Q_{sign} *defined on* $H^\bullet(X)$.

PROOF. Decompose $H^\bullet(X)$ according to the degree is smaller, equal to, or larger than m to obtain $H^{<m} \oplus H^m \oplus H^{>m}$. Then H^m and $H^{<m} \oplus H^{>m}$ are orthogonal with respect to Q_{sign}. Hence it is enough to show that the signature of $H^{<m} \oplus H^{>m}$ is 0. By an algebraic argument, we find this claim from the fact that the restriction of Q_{sign} to $H^{<m}$, resp. $H^{>m}$, vanishes identically. □

REMARK 2.52.

(1) When m is odd, $\sqrt{-1}\, Q_{\text{sign}}$ extends to $H^\bullet(X) \otimes \mathbb{C}$ as a Hermitian form. We can define $\operatorname{sign} X$ by its signature. But it is always 0.
(2) For a self diffeomorphism $g : X \to X$ of X, denote by $(g^{-1})^*$ the action on the de Rham cohomology, which is induced by g^{-1}. Take the tensor product with $\mathbb{C}$ and consider the (generalized) eigenspace decomposition of this linear action. On each (generalized) eigenspace, extend Q_{sign} or $\sqrt{-1}\, Q_{\text{sign}}$ as a Hermitian form. Then we can consider its signature. Denote by $\operatorname{sign}_g X$ the sum of these signatures multiplied by the corresponding eigenvalues. It is a generalization of $\operatorname{sign} X$.
(3) For a local system θ with the structure group being the unitary group, it is possible to perform a similar construction. Construct Q_{sign} or $\sqrt{-1}\, Q_{\text{sign}}$ as a Hermitian form. Then its signature $\operatorname{sign}_\theta X$ is defined.
(4) Even in the case that m is odd, $\operatorname{sign}_g X$, resp. $\operatorname{sign}_\theta X$, may not be 0 for a self diffeomorphism g, resp. a local system θ with the structure group being the unitary group.

Up to now, we discuss the story depending only on purely topological information. Pick an arbitrary Riemannian metric on X, then we can find a relation of $\operatorname{sign} X$ with a certain operator of Dirac type.

In general, fixing a Euclidean [Hermitian] metric on a real [complex] vector space equipped with a symmetric [Hermitian] bilinear form, we can express the symmetric [Hermitian] bilinear form by a symmetric [Hermitian] endomorphism. Then the signature of the symmetric [Hermitian] bilinear form coincides with the difference of the number of positive eigenvalues and negative eigenvalues of the corresponding endomorphism.

We would like to apply this idea to $\operatorname{Ker} D^2$ in §4.1.

Fix a Riemannian metric on X and consider the real vector bundle $W = \bigwedge^\bullet T^*X$ in §4.1. Define a section of

$$\tilde{*} \bigoplus_k \operatorname{Hom}(\bigwedge^k T^*X, \bigwedge^{n-k} T^*X)$$

so that it satisfies

$$(-1)^{\frac{1}{2}l(l+1)} u \wedge v = \langle u, \tilde{*} v \rangle d\,\mathrm{vol}$$

pointwisely for a differential k-form u and a differential l-form v with $k + l = n$. Here $d\,\mathrm{vol}$ is the differential n-form determined by the Riemannian metric and the orientation of X. Denote by vol the section of $Cl(T^*X)$ corresponding to $d\,\mathrm{vol}$.

LEMMA 2.53.

(1) $\tilde{*}$ *coincides with the Clifford multiplication by* vol *on the real Clifford module* W *over* (X, T^*X).

(2) $D = d + d^*$ *and* vol *satisfy the following relation:*

$$D(\mathrm{vol}) + (-1)^n (\mathrm{vol}) D = 0\,.$$

The above lemma holds, even when n is not an even number.

We postpone the proof of this lemma. Assuming this lemma, we explain the relation between $\operatorname{sign} X$ and the dimension of solutions of a certain operator of Dirac type.

In the previous section, we equipped the real Clifford module $W = \bigwedge^\bullet T^*X$ with $\mathbb{Z}_2$-grading by the parity of degree of differential forms.

Recall that we assume that X is oriented and of even dimension. Then Lemma 2.16 implies that we can introduce a $\mathbb{Z}_2$-grading on W by the eigenspace decomposition of the action by vol. But this differs from the $\mathbb{Z}_2$-grading by the degree of differential forms.

When the dimension of X is $n = 2m$, we have $(\mathrm{vol})^2 = (-1)^{n(n+1)/2} = (-1)^m$ as the action of the Clifford multiplication.

(1) When m is even, the eigenvalues of vol are ± 1. Define the decomposition $W = W^0 \oplus W^1$ so that

$$(-1)^{\frac{1}{2}m(m+1)}\,\mathrm{vol} = \epsilon_W\,.$$

(2) When m is odd, the eigenvalues of vol are $\pm\sqrt{-1}$. Define the decomposition $W = W^0 \oplus W^1$ so that

$$\sqrt{-1}\,(-1)^{\frac{1}{2}m(m+1)}\,\mathrm{vol} = \epsilon_W\,.$$

By Lemma 2.53(2) (since we assume that n is even), D is of degree 1 with respect to this $\mathbb{Z}_2$-grading and is regarded as an operator of Dirac type. In particular, vol acts on $\operatorname{Ker} D^2$.

When m is even, by Lemma 2.53(1), the $\mathbb{Z}_2$-grading and Q_{sign} correspond to each other with the help of the Riemannian metric. In particular, Q_{sign} is positive definite on W^0 and negative definite on W^1. Hence we obtain the following proposition.

PROPOSITION 2.54. *Let X be a $2m$-dimensional oriented closed manifold. We assume that m is even. Then*

$$\dim \operatorname{Ker}(D^2|\Gamma(W^0)) - \dim \operatorname{Ker}(D^2|\Gamma(W^1))$$

coincides with the signature $\operatorname{sign} X$ *of* X.

REMARK 2.55.

(1) Since 0 is not an eigenvalue of vol, we find that the bilinear form Q_{sign} is non-degenerate. This is one way of presenting the Poincaré duality. This argument can be extended to oriented closed manifolds of any dimension.
(2) A similar proposition also holds in the case that m is odd.

In the remaining part of this section, we prove Lemma 2.53.

First of all, we discuss how the action of vol is. We consider this question locally.

Let $(e_1, e_2, \ldots, e_n)$ be a local orthonormal frame of X compatible with the given orientation. Then we have

$$\text{vol}\cdot = (e_1^{\wedge} - e_1^{\lrcorner})(e_2^{\wedge} - e_2^{\lrcorner}) \cdots (e_n^{\wedge} - e_n^{\lrcorner}).$$

From this definition, we find that

$$\text{vol}((-e_n) \wedge (-e_{n-1}) \wedge \cdots \wedge (-e_{k+1})) = e_1 \wedge e_2 \wedge \cdots \wedge e_k.$$

When $k + l = n$, we can rephrase it in the following way:

$$(-1)^{\frac{1}{2}l(l+1)} \text{vol}(e_{k+1} \wedge e_{k+2} \wedge \cdots \wedge e_n) = e_1 \wedge e_2 \wedge \cdots \wedge e_k .$$

On the other hand, we define the **Hodge's $*$-operator** as the linear operator $*$ acting on W such that $\langle u, v\rangle d\,\text{vol} = u \wedge *v$. (Since the exterior product is a non-degenerate bilinear form, we clearly find the existence and uniqueness of $*$.) For example, it is given explicitly by

$$*(e_1 \wedge e_2 \wedge \cdots \wedge e_k) = e_{k+1} \wedge e_{k+2} \wedge \cdots \wedge e_n.$$

Since it holds for any orthonormal basis compatible with the orientation, we find

$$(-1)^{l(l+1)/2}\,\text{vol} = *^{-1}$$

on $\bigwedge^l T^*X$ by comparison. On the other hand, rephrasing the definition of $*$ as $\langle u, *^{-1}v\rangle d\,\mathrm{vol} = u \wedge v$, we find that

$$\tilde{*} = (-1)^{l(l+1)/2} *^{-1}$$

on $\bigwedge^l T^*X$. From this, we obtain vol $= \tilde{*}$ in the first half of the lemma.

To give an explicit expression of d^*, it is convenient to use Hodge's $*$ operator. Since X is an oriented closed manifold, we have

$$0 = \int_X d(s_0 \wedge *s_1) = \int_X (ds_0) \wedge *s_1 + (-1)^{\deg s_0} \int_X s_0 \wedge d(*s_1)$$

for differential forms s_0, s_1, hence we obtain

$$\int_X (ds_0) \wedge *s_1 = -(-1)^{\deg s_0} \int_X s_0 \wedge *(*^{-1} d(*s_1)).$$

This implies that

$$d^* = -(-1)^l *^{-1} d*$$

on $\Omega^k(X)$, if $k + l = n$. Rewriting this equality using vol, we easily check that

$$(\mathrm{vol})d^* + (-1)^n d(\mathrm{vol}) = 0, \quad (\mathrm{vol})d + (-1)^n d^*(\mathrm{vol}) = 0.$$

The second half of the lemma follows from this fact immediately.

REMARK 2.56. The operator $D(\mathrm{vol}) + (-1)^n(\mathrm{vol})D$ is a linear differential operator of order at most 1, and we can show that its principal symbol is 0 from the fact that D is an operator of Dirac type only. In particular, if $D = d + d^*$, the second half of Lemma 2.53 claims that this is identically 0 as an operator.

SUMMARY.

2.1 We can define the integration along fibers for differential forms with compact support along fibers.

2.2 Any manifold can be embedded in the Euclidean space as a closed submanifold. Any vector bundle can be embedded in a trivial vector bundle as a vector sub-bundle.

2.3 For a Clifford module on the cotangent bundle, we can define the operator of Dirac type.

2.4 Geometric objects such as the de Rham cohomology, the Dolbeault cohomology, etc., can be expressed as the kernel of the operator of Dirac type on suitable Clifford modules.

CHAPTER 3

Index and Its Localization

In this chapter, we explain the method of "localization".

First of all, in § 1, we define the index for $\mathbb{Z}_2$-graded Clifford modules on closed manifolds. In order to study the behavior of the index under cut-and-paste construction, we need to extend the definition so that it can be applied to open manifolds. In such a case, we assume that the Clifford modules have compact support. In § 2, we use the localization of eigenfunctions for the Laplace-type operator to give the definition to the extended case. The proof of localization will be given in Chapter 4. Once we are able to arrive at the appropriate definition, the properties, which the index enjoys as the reflection of the localization principle, can be formulated as the "Excision Theorem".

When we study the index by cutting-and-pasting manifolds, the de Rham complex (with coefficients in the orientation local system) on the Euclidean space plays a fundamental role.

In § 5, we define the supersymmetric harmonic oscillator as the compactly supported $\mathbb{Z}_2$-graded Clifford module, which is closely related to the de Rham complex mentioned above.

1. Definition of the Index of Operators of Dirac Type on Closed Manifolds

Let $W = W^0 \oplus W^1$ be a $\mathbb{Z}_2$-graded Clifford module on a closed manifold X. W is a topological object. (Precisely speaking, the associated principal bundle of W is a topological object. For each Riemannian metric on X, the vector bundle W is defined.) From now on, we fix a Riemannian metric on X.

Denote by D an operator of Dirac type on W. The fundamental property of D or D^2 is the following finiteness.

THEOREM 3.1 (cf. Theorem 5.43). *Let X be a closed manifold. Then the eigenvalues of D^2 are non-negative real numbers. For each*

$\lambda \geq 0$, the number of eigenvalues of D^2, which are not greater than λ, is finite counting with multiplicity.

We state a generalization of the above theorem in the next subsection. We will prove them in Chapter 4. Let us assume Theorem 3.1 for the moment. Then, in particular, the space of solution of $Ds = 0$ is finite dimensional. Hence the following definition makes sense.

DEFINITION 3.2. We define the index $\operatorname{ind} D$ of the operator D of Dirac type on closed manifold X as follows.

$$\begin{aligned}\operatorname{ind} D &= \dim \operatorname{Ker}(D : \Gamma(W^0) \to \Gamma(W^1)) \\ &\quad - \dim \operatorname{Ker}(D : \Gamma(W^1) \to \Gamma(W^0)).\end{aligned}$$

Since D is formally self-adjoint, the integration by parts insures $\operatorname{Ker} D = \operatorname{Ker} D^2$. Hence we obtain

LEMMA 3.3. *The index of the operator D of Dirac type on a closed manifold X is equal to the following number:*

$$\begin{aligned}\operatorname{ind} D &= \dim \operatorname{Ker}(D^2 : \Gamma(W^0) \to \Gamma(W^0)) \\ &\quad - \dim \operatorname{Ker}(D^2 : \Gamma(W^1) \to \Gamma(W^1)).\end{aligned}$$

For each $i = 0, 1$, we denote by $E_\lambda(D^2|\Gamma(W^i))$ (resp. $E_{\leq\lambda}(D^2|\Gamma(W^i))$) the finite dimensional vector space spanned by eigenfunctions with eigenvalues $\lambda \geq 0$ (resp. or at most λ) of $D^2|\Gamma(W^i)$. Then for $\lambda > 0$, D induces a linear isomorphism

$$D : \ E_\lambda(D^2|\Gamma(W^0)) \to E_\lambda(D^2|\Gamma(W^1)), \qquad (\lambda > 0).$$

Therefore, we obtain the following expression of the index.

LEMMA 3.4. *Let X be a closed Riemannian manifold. Then, for each $\bar{\lambda} \geq 0$, we find the following:*

$$\operatorname{ind} D = \dim E_{\leq\bar{\lambda}}(D^2|\Gamma(W^0)) - \dim E_{\leq\bar{\lambda}}(D^2|\Gamma(W^1)).$$

This is because the eigenspaces with non-zero eigenvalues cancel each other.

The most fundamental property of the index is that the index is determined by topological information.

THEOREM 3.5. *For a Clifford module W on a closed manifold X, $\operatorname{ind} D$ of the operator of Dirac type does not depend on the choice of D and a Riemannian metric on X, and is determined only by W.*

The proof is given by showing that the eigenvalues of D^2 change continuously when we change the choices of D and Riemannian metrics on X continuously.

Using Lemma 3.4 and Theorem 5.31, which will be proved in Chapter 5, §5, we can find that the index is invariant under these continuous deformations. (The point of the proof is that the right hand side of the formula in Lemma does not change under small deformations of the choices of D and Riemannian metrics, when the number λ in Lemma above is chosen as a non-eigenvalue.) We will give, in the next section, the proof in more general cases.

2. Definition of the Index of Operators of Dirac Type on Open Manifolds

2.1. Clifford Modules with Compact Support. In this section, X denotes a manifold, which is not necessarily closed.

We would like to consider additional structures, which forces restriction to the behavior of the operator of Dirac type at the end of an open manifold. Here, an "end" means the complement of a compact subset. It is the structure which makes it possible to study the problem essentially on compact subsets. The idea is to consider two Clifford modules, which are isomorphic at the end of the manifold. At this point, we would like to define a kind of "difference" of two Clifford modules. Then, roughly speaking, the two Clifford modules cancel each other at the end of the manifold by the existence of the isomorphism, and the "difference" is expected to "localize" on the compact subset.

The next definition is a preparation to formulate this idea. (However, this definition is only for explanation of the idea and we do not use it later.)

Generally, let T be a Euclidean vector bundle on X and W a $\mathbb{Z}_2$-graded Clifford module over T.

DEFINITION 3.6. A Hermitian vector bundle W on X is called a $\mathbb{Z}_2 \times \mathbb{Z}_2$**-graded Clifford module** if the following conditions are satisfied.

(1) A direct sum decomposition $W = W^{00} \oplus W^{01} \oplus W^{10} \oplus W^{11}$ is given.
(2) Each of $W^{0\bullet} = W^{00} \oplus W^{01}$ and $W^{1\bullet} = W^{10} \oplus W^{11}$ is equipped with a structure of $\mathbb{Z}_2$-graded Clifford module.

(More precisely, it must be called a $\mathbb{Z}_2 \times \mathbb{Z}_2$-graded Clifford module with the degree of the Clifford multiplication being $(0,1)$. But this terminology is too long and we just call it a $\mathbb{Z}_2 \times \mathbb{Z}_2$-graded Clifford module.)

When a pair of two $\mathbb{Z}_2$-graded Clifford modules $W^{0\bullet}, W^{1\bullet}$ is given, we define the Clifford multiplication on $W = W^{0\bullet} \oplus W^{1\bullet}$ by $c_W = c_{W^{0\bullet}} - c_{W^{1\bullet}}$. Then it becomes a $\mathbb{Z}_2 \times \mathbb{Z}_2$-graded Clifford module. Conversely, any $\mathbb{Z}_2 \times \mathbb{Z}_2$-graded Clifford module is realized in this way.

We will regard any $\mathbb{Z}_2$-graded Clifford module $W^0 \oplus W^1$as a special kind of $\mathbb{Z}_2 \times \mathbb{Z}_2$-graded Clifford module by setting $W^{0i} = W^i$, $W^{1i} = 0$.

A typical example of $\mathbb{Z}_2 \times \mathbb{Z}_2$-graded Clifford module is the tensor product of a $\mathbb{Z}_2$-graded Clifford module S and any $\mathbb{Z}_2$-graded Hermitian vector bundle F, i.e., $W = F \otimes S$. We will mainly discuss the Clifford modules of this type.

$$W^{00} = F^0 \otimes S^0, \qquad W^{10} = F^1 \otimes S^0 . \tag{3.1}$$

$$W^{01} = F^0 \otimes S^1, \qquad W^{11} = F^1 \otimes S^1 . \tag{3.2}$$

$$c_W = \epsilon_F \otimes c_S . \tag{3.3}$$

REMARK 3.7. By Proposition 2.39 (1), if X is of even dimension and carries a $Spin^c$-structure, any $\mathbb{Z}_2$-graded Clifford module is expressed as the $\mathbb{Z}_2 \times \mathbb{Z}_2$-graded Clifford module as above with the spinor bundle S.

Next, we discuss the isomorphism between two Clifford modules at the end of X. Giving the data not only on the end but the whole X, it is convenient that we need not to separate the discussion on the compact part and at the end. Thus we firstly consider the homomorphism on the whole X and introduce the following definition.

DEFINITION 3.8. Let $\operatorname{Herm} W$ be the vector subbundle of $\operatorname{End} W$, which consists of Hermitian transformations of W. We define the vector subbundle $H_{\mathbb{Z}_2 \times \mathbb{Z}_2}$ of $\operatorname{Herm} W$ consisting of all h as follows.

- The degree of h is $(1,0)$.
- h anti-commutes with the Clifford multiplication: $hc_W + c_W h = 0$.

REMARK 3.9.

(1) $H_{\mathbb{Z}_2\times\mathbb{Z}_2}$ is identified with the set of all homomorphisms $\underline{h}_W$ from $W^{0\bullet}$ to $W^{1\bullet}$, which preserve the Clifford multiplication. In fact, the identification is given by the following correspondence:

$$\underline{h}_W \mapsto h = \begin{pmatrix} 0 & \underline{h}_W^* \\ \underline{h}_W & 0 \end{pmatrix}.$$

(2) We assume that X is of even dimension and carries a $Spin^c$-structure and denote by S the spinor bundle. Then for $W = F \otimes S$, it is easy to see that $H_{\mathbb{Z}_2\times\mathbb{Z}_2}$ is isomorphic to $\operatorname{Hom}(F^0, F^1)$. We explain the above correspondence in a different order. Firstly, $\operatorname{Hom}(F^0, F^1)$ is in one-to-one correspondence to the set $\operatorname{Herm} F$ of all Hermitian endomorphisms of F of degree 1 by

$$\underline{h}_F \mapsto h_F = \begin{pmatrix} 0 & \underline{h}_F^* \\ \underline{h}_F & 0 \end{pmatrix}.$$

In addition to this, we assign

$$h_F \mapsto h = h_F \otimes \mathrm{id}_S,$$

which is an element of $H_{\mathbb{Z}_2\times\mathbb{Z}_2}$. This is the desired isomorphism.

We expect that there occurs a cancellation between $W^{0\bullet}$ and $W^{1\bullet}$ on the part of X, where h is isomorphic.

DEFINITION 3.10. We define the **support** of h as the set of all points where h is not isomorphic and denote it by $\operatorname{supp} h$. $\operatorname{supp} h$ is a closed set.

Now the desired formulation is as follows.

DEFINITION 3.11. A pair $(W, [h])$ is called a $\mathbb{Z}_2\times\mathbb{Z}_2$**-graded Clifford module with compact support** when it satisfies the following conditions:

(1) W is a $\mathbb{Z}_2 \times \mathbb{Z}_2$-graded Clifford module.
(2) h is a section of $H_{\mathbb{Z}_2\times\mathbb{Z}_2}$ and its support $\operatorname{supp} h$ is a compact subset of X.
(3) $[h]$ is the homotopy class of h among those satisfying the conditions above.

In particular, when X itself is compact, the pair $(W, [h])$ is a $\mathbb{Z}_2 \times \mathbb{Z}_2$-graded Clifford module with compact support for any h. Moreover, the possibility of homotopy classes is unique. Hence we identify $(W, [h])$ and W in this case.

REMARK 3.12.

(1) If $(W, [h])$ is a $\mathbb{Z}_2 \times \mathbb{Z}_2$-graded Clifford module with compact support on a non-compact manifold X, all the ranks of four W^{ij} must be the same.
(2) Let D be any operator of Dirac type. Then $Dh + hD$ is formally a differential operator of order 1. Since h is an element of $H_{\mathbb{Z}_2 \times \mathbb{Z}_2}$, the terms involving differentiation cancel each other and it turns out a Hermitian endomorphism of the vector bundle W. This is the most important point in the definition above.

At first glance, the definition of $\mathbb{Z}_2 \times \mathbb{Z}_2$-graded Clifford modules looks complicated. It, however, faithfully represents the motivation to define the difference between two Clifford modules. But it is more suitable, for later purposes, to find simpler and more general objects.

DEFINITION 3.13. Let W be a $\mathbb{Z}_2$-graded Clifford module over a Euclidean bundle T.

(1) We define a vector subbundle $H_{\mathbb{Z}_2}$ of $\operatorname{Herm} W$ as the set of all h of degree 1, which anti-commute with the Clifford multiplication: $hc_W + c_W h = 0$.
(2) $(W, [h])$ is called a $\mathbb{Z}_2$-**graded Clifford module with compact support** over (X, T), if h is a section of $H_{\mathbb{Z}_2 \times \mathbb{Z}_2}$ and the support of h is compact. Denote by $[h]$ the homotopy class of h satisfying these conditions.

A $\mathbb{Z}_2 \times \mathbb{Z}_2$-graded Clifford module with compact support is regarded as a $\mathbb{Z}_2$-graded Clifford module with compact support by using the total degree. The central objects we discuss in this book are these $\mathbb{Z}_2$-graded Clifford modules with compact supports.

The supersymmetric harmonic oscillator, which appears as an important example later, is $\mathbb{Z}_2$-graded, but does not carry a structure of $\mathbb{Z}_2 \times \mathbb{Z}_2$-grading.

DEFINITION 3.14. For simplicity, we call, from now on, a $\mathbb{Z}_2$-graded Clifford module with compact support a **pair**$(W, [h])$.

We fix a Riemannian metric on X and consider $T = T^*X$. Then we call a pair $(W, [h])$ over (X, T^*) a **$\mathbb{Z}_2$-graded Clifford module on X with compact support**, or a pair on X.

REMARK 3.15.

(1) We explained the meaning of $H_{\mathbb{Z}_2}$ as an extension of $H_{\mathbb{Z}_2\times\mathbb{Z}_2}$. We can give another explanation by considering the "product" of Dirac operators. See § 4.
(2) Using an appropriate fiber bundle, we can define the notion of a "pair on X" as a purely topological object regardless of a Riemannian metric. We, however, do not pursue it, since our study of the index is not affected by deforming h up to homotopy (keeping the conditions in the definition) as we will see.

Let $(W, [h])$ be a pair on X and D an operator of Dirac type on W. We would like to define the index of $D_t = D + th$ for a sufficiently large $t > 0$.

In order to define it, we need to choose a Riemannian metric on X and an operator D of Dirac type, which behave nicely (in some sense) at the end of X, and also introduce suitable function spaces, on which D_t acts. Moreover, we would like to establish that the index does not depend on these auxiliary tools and only depends on the topological data, i.e., the pair $(W, [h])$ (especially, the homotopy class of h) finally. In this book, we use "localization of eigenfunctions of operators of Laplace type" to pursue this program.

2.2. Riemannian Metrics on Open Manifolds. We discuss how to deal with ends of open manifolds in the following order:

(1) We consider the case that X is diffeomorphic to the interior of a compact manifold with boundary. We consider Riemannian metrics on X, which are "cylindrical" at the end (§ 2.3).
(2) In (1), most of the arguments work for Riemannian metrics with milder conditions at the end, i.e. Riemannian metrics, which are isomorphic to the Euclidean space at the end. The topological constraint for the end of X is the same as the cylindrical case. But it is also necessary to study such Riemannian metrics in applications (§ 2.4).

(3) For a general X, we prove topological invariance of the index by studying the "excision theorem", which is a consequence of "localization" (§ 3).

2.3. Manifolds with Cylindrical Ends. Let us consider the case that X is diffeomorphic to the interior of a compact manifold $\bar{K}$ with boundary.

Denote by Z the boundary of $\bar{K}$. Then X is obtained by gluing K_0, which is diffeomorphic to $\bar{K}$ and $Z \times [0, \infty)$ along the boundary Z and $Z \times \{0\}$:

$$X = K_0 \cup Z \times [0, \infty).$$

In this section, we fix a Riemannian metric on X, which is the product of some Riemannian metric on Z and the standard Riemannian metric on $[0, \infty)$ on $Z \times [0, \infty)$. We call the part $Z \times [0, \infty)$ **cylinder**.

On $Z \times \mathbb{R}$, $\mathbb{R}$ acts on the right factor additively. We call this action the **translation**. A Clifford module on $Z \times \mathbb{R}$ is called translationally invariant, if the $\mathbb{R}$-action lifts to the Clifford module and preserves the structure of the Clifford module (the Hermitian inner product, the Clifford multiplication).

If a Clifford module on the cylinder $Z \times [0, \infty)$ is the restriction of a translationally invariant Clifford module on $Z \times \mathbb{R}$, we also call it translationally invariant. We use similar terminology for other structures, although we do not define them.

Let $(W, [h])$ be a translationally invariant pair on the cylinder (more precisely, W and a representative h of $[h]$ are translationally invariant). Pick an operator D of Dirac type, which is translationally invariant on the cylinder.

What we would like to study is $D_t = D + th$ for a sufficiently large $t > 0$. This operator is of degree 1 and interchanges sections of W^0 and sections of W^1. We study the following operator of degree 0:

$$D_t^2 = D^2 + t(Dh + hD) + t^2h^2.$$

D_t^2 preserves $\Gamma(W^0)$ and $\Gamma(W^1)$, respectively.

Set $V_t = t(Dh + hD) + t^2h^2$. Then V_t is a Hermitian endomorphism of W. W and $D_t^2 = D^2 + V_t$ acting on $\Gamma(W)$ satisfy the following conditions:

(1) W is a Clifford module on X, and D is an operator of Dirac type on W, V_t is a Hermitian endomorphism of W.

(2) W, D, V_t are translationally invariant on the cylinder at the end of X.

(3) For any real number λ, we choose t sufficiently large so that $V_t \geq \lambda$ on the cylinder.

For an operator D of Dirac type and a Hermitian endomorphism V_t, $D^2 + V_t$ is a variant of an operator of Laplace type, which we will define in Definition 5.1 in Chapter 5.

We discuss eigenvalues of D^2+V_t. To do so, we need to introduce suitable function spaces for eigenfunctions.

DEFINITION 3.16. For a Hermitian vector bundle W on $X = K_0 \cup Z \times [0, \infty)$, we define $B_X(W)$ the set of all smooth sections s satisfying the following condition:

$$\liminf_{l \to \infty} \int_{Z \times \{l\}} (|Ds|^2 + |hs|^2) = 0.$$

From now on, we consider eigenfunctions belonging to $B_X(W)$. Let s be an eigenfunction of D_t^2, which belongs to $B_X(W)$. Using the "exponential decay" (Lemma 5.9, Chapter 5, and its proof) and "a-priori estimate" (Proposition 5.20), which we will explain in Chapter 5, we can find that $D_t s$ also belongs to $B_X(W)$. Hence, if λ is not 0, the set of eigenfunctions of D_t^2 belonging to $B_X(W^0)$ and the set of eigenfunctions of D_t^2 belonging to $B_X(W^1)$ are isomorphic through D_t.

We will prove the following theorem in Chapter 5 (cf. Corollary 5.28).

THEOREM 3.17. *Let $X = K_0 \cup Z \times [0, \infty)$ be a Riemannian manifold with cylindrical end, and $(W, [h])$ a pair on X. We suppose that W, D, h are translationally invariant on the cylinder and pick a positive number $\bar{\lambda}, t > 0$ such that $V_t = t(hD + Dh) + t^2 h^2 > \bar{\lambda}$ holds on the cylinder. Then there are only finitely many eigenvalues, counting with multiplicity, of D_t^2 less than or equal to $\bar{\lambda}$.*

Therefore, for D_t satisfying the conditions of this theorem, we can define the index $\operatorname{ind} D_t$ by the same formula as in Lemma 3.3 or Lemma 3.4. We also have

LEMMA 3.18. *For any $0 \leq \lambda \leq \bar{\lambda}$, the following holds:*

$$\operatorname{ind} D_t = \dim E_{\leq \lambda}(D_t^2 | B_X(W^0)) - \dim E_{\leq \lambda}(D_t^2 | B_X(W^1)).$$

Next, we would like to claim that this definition does not depend on auxiliary data.

THEOREM 3.19 (cf. Theorem 5.38, Chapter 5). *Let* $X = K_0 \cup Z \times [0, \infty)$ *be a Riemannian manifold with cylindrical end, and* $(W, [h])$ *a pair on* X. *We suppose that* W, D, h *are translationally invariant on the cylinder and pick a positive number* $\bar{\lambda}, t > 0$ *such that* $V_t = t(hD + Dh) + t^2h^2 > \bar{\lambda}$ *holds on the cylinder. Assume that* $\mu < \bar{\lambda}$ *is not an eigenvalue of* D_t^2. *Then the number of eigenvalues (on both* $B_X(W^0)$ *and* $B_X(W^1)$*) less than or equal to* μ *is constant, when we continuously deform the Riemannian metric on* X *and* D, h, t *slightly so that the following conditions are satisfied:*

- *The Riemannian metric is cylindrical on* $Z \times [0, \infty)$, *and* D, h *are translationally invariant on the cylinder.*
- *The estimate* $V_t > \bar{\lambda}$ *holds on the cylinder.*

Here, "continuous deformation" means the following. At each point, we pick local coordinates and a trivialization of W on its neighborhood. Using them, we express D, V_t and the Riemannian metric on X in coordinates. Then their coefficients are continuous on the neighborhood and the parameter space. (More precisely, as we noticed at the end of Chapter 2, § 3, the notion of Clifford modules on X depends on the Riemannian metric on X, thus W itself changes continuously as vector bundles. A vector bundle corresponding to W is defined on the product of X and the deformation parameter space.)

From this theorem, we can prove that the index is determined only by the topological data, i.e., the pair $(W, [h])$.

THEOREM 3.20. *Suppose that* X *is diffeomorphic to the interior of a compact manifold with boundary. Let* $(W, [h])$ *be a pair on* X. *Pick an arbitrary Riemannian metric on* X, *which is cylindrical at the end and deform* h *by homotopy (keeping the support compact) so that* h *is translationally invariant on the cylinder. Then, for a translationally invariant* D *and a sufficiently large* $t > 0$, $\operatorname{ind} D_t$ *is independent of the choice of a Riemannian metric and* h, D, t.

PROOF. It is enough to show that $\operatorname{ind} D_t$ is invariant, when we change, continuously in some parameter ω, the Riemannian metric and h, D, t with keeping the conditions in the theorem. Pick $\bar{\lambda} > 0$ so that $t(hD + Dh) + t^2h^2 > \bar{\lambda}$ holds at $\omega = \omega_0$. Since there are only finitely many eigenvalues of D_t^2 less than or equal to $\bar{\lambda}$, we can choose $0 < \mu < \bar{\lambda}$, which is not any eigenvalue. By Theorem 3.19, the number of eigenvalues less than or equal to μ does not change for both $B_X(W^0)$ and $B_X(W^1)$, whenever the parameter ω is sufficiently close to ω_0. Then the conclusion follows from Lemma 3.18. □

Based on this definition, we set

DEFINITION 3.21. $\mathrm{ind}(W, [h]) = \mathrm{ind}\, D_t$.

2.4. Ends with More General Riemannian Metrics. In the above formulation, we applied results in Chapter 5 to operators of Laplace type with translationally invariant data on the cylinder. For X diffeomorphic to the interior of a compact manifold with boundary, we can also apply results in Chapter 5 to the Riemannian metric not necessarily with cylindrical end.

In fact, we can extend the argument to a Riemannian manifold X satisfying the following condition (R):

Let $f : X \to \mathbb{R}$ be a proper smooth map satisfying the following.

Write $K_0 = f^{-1}((-\infty, 0])$ and set $B_r^\circ(p)$ the set of all points, the distance from p to which is less than r.

ASSUMPTION 3.22 (Condition on Riemannian metrics (R)).

- K_0 is compact.
- For any point p in X, there is a point q in K_0 such that $B_1^\circ(p)$ is isomorphic to $B_1^\circ(q)$ as Riemannian manifolds, i.e., isometric.
- On $X \setminus K_0$, $0 < \inf |\nabla f|$ and $\sup |\nabla f| < +\infty$ hold. Here, ∇f is the gradient vector with respect to the Riemannian metric, which we consider.

The first two conditions correspond to the translational invariance. The last means that, if X is not a closed manifold, X expands to the infinity and f almost faithfully reflects the degree of expansion. Other than Riemannian manifolds with cylindrical ends, there are following examples of Riemannian manifolds satisfying the condition (R).

EXAMPLE 3.23.

(1) For a closed Riemannian manifold Z and a positive integer r, X, the end of which is isomorphic to the end of $Z \times \mathbb{R}^r$, satisfies the condition (R). (Define f by using the distance from the origin in $\mathbb{R}^r$.)

(2) If X, X' satisfy the condtion (R), then the Riemannian product $X \times X'$ also satisfies the condition (R). (Use the sum of f for X and f' for X'.)

We present here examples of Riemannian manifolds satisfying the condition (R), which are necessary to establish the index theorem.

For a Riemannian manifold X with cylindrical end and ν a real vector bundle on X, we define the Riemannian metric g_ν on the total space of ν in the following way.

Firstly, we consider the case that X is a point and ν is identified with a Euclidean space $\mathbb{R}^m$. We pick a Riemannian metric $g_{\mathbb{R}^m}$ on $\mathbb{R}^m$ and a smooth function $f_{\mathbb{R}^m}$, which satisfy the following conditions. Denote by $r : \mathbb{R}^m \to [0, \infty)$ the distance from the origin with respect to the standard Riemannian metric on $\mathbb{R}^m$. The orthogonal group $O(m)$ acts on $\mathbb{R}^m$ preserving the Riemannian metric and fixing the origin.

- $g_{\mathbb{R}^m}$ and $f_{\mathbb{R}^m}$ are $O(m)$-invariant. In particular, f is a function on the distance r from the origin.
- $g_{\mathbb{R}^m}$ coincides with the standard Riemannian metric on $\mathbb{R}^m$ in the region $r < 1$, and $g_{\mathbb{R}^m}$ is isometric to the product $S^{m-1} \times [0, \infty)$ in the region $r > 2$.
- f coincides with $r^2 - 2$ in the region $r < 1$, is monotone increasing in r and coincides with the $[0, \infty)$-component of $S^{m-1} \times [0, \infty)$ in the region $r > 2$.

In this case, $g_{\mathbb{R}^m}$ is a Riemannian metric on $\mathbb{R}^m$ with cylindrical end.

For a general X, we define g_ν as follows.

(1) Fix a Riemannian inner product (Euclidean metric) on ν, which is translationally invariant on the cylinder at the end of X.
(2) Define the Riemannian metric on each fiber $(\nu)_x$ of ν in such a way that it is isomorphic to $g_{\mathbb{R}^m}$ via the linear isomorphism between $(\nu)_x$ and $\mathbb{R}^m$ preserving the inner product. (Since $g_{\mathbb{R}^m}$ is $O(m)$-invariant, it does not depend on the choice of the linear isomorphism preserving the inner product.)
(3) Fix a splitting of the natural exact sequence

$$0 \to T_{\text{fiber}}\nu \to T\nu \to \pi^*TX \to 0$$

which is translationally invariant on the cylinder. Here, $T_{\text{fiber}}\nu$ is the vector bundle on ν consisting of all tangent vectors along fibers of ν.
(4) The above splitting induces an isomorphism $T\nu \cong T_{\text{fiber}}\nu \oplus \pi^*TX$. Using this isomorphism, we define the metric g_ν on the left hand side by the metric on the right hand side.

LEMMA 3.24. *g_ν satisfies the condition* (R).

PROOF. Identifying each fiber of ν with $\mathbb{R}^m$ via a linear isomorphism preserving the inner products, we regard $f_{\mathbb{R}^m}$ as a function on each fiber. Since $f_{\mathbb{R}^m}$ is $O(m)$-invariant, this function does not depend on the choice of the linear isomorphism preserving the inner products. These functions for all fibers gives a function on the total space of ν, which we denote by f_0. We also pick a smooth function f on X, which coincides with the $[0,\infty)$-component on the cylinder $Z \times [0,\infty)$ and is negative on a relatively compact part in the complement of the cylinder. If necessary, we replace f by $f - c$ for some positive constant c and assume that the Riemannian metric on X satisfies the condition (R) with respect to f. Denote by π the projection from ν to X. Then for the function $f_\nu = f \circ \pi + f_0$ on the total space of ν, g_ν satisfies the condition (R). We leave the details to the reader. □

For D, h, we assume the following condition.

ASSUMPTION 3.25 (Condition (D) on D, h).

- For each point p in X, there exists a point q in K_0 such that the restriction of D to $B_1^\circ(p)$ is isomorphic to the restriction of D to $B_1^\circ(q)$.
- For any positive number λ, if the real number t is sufficiently large, the following condition holds on each point in the cylinder:

$$t(Dh + hD) + t^2h^2 > \lambda.$$

Note that we do not assume translational invariance for h, while we assume it for D.

EXAMPLE 3.26. Euclidean space satisfies the condition (R). The supersymmetric harmonic oscillator on the Euclidean space, which is defined in § 5 below, is not translationally invariant, but satisfies Condition (D).

REMARK 3.27. In the argument in Chapter 5, we can easily replace "isometric" in the first half of Condition (R) by a milder condition expressed by an inequality. It is also the case for the first half of Condition (D). In the proof of the index theorem in this book, we can choose a suitable Riemannian metric and D according to our purpose. It is, however, necessary to study the problem under such a milder condition, when a Riemannian metric and/or D arises in geometric situations (e.g., instantons or Einstein metrics on open manifolds).

We must fix the function spaces in order to study eigenvalues, eigenfunctions. We adopt the definition of B_X in Definition 3.16 with the integration over $Z \times \{l\}$ replaced by the integration over $f^{-1}(l)$.

Under this setting, a claim corresponding to Theorem 3.19 holds without change. Hence, we can define $\operatorname{ind} D_t$. In the next section, we will explain that this does not depend on auxiliary data.

Under Condition (R), we do not require any condition on h (hence on $[V_t]$) corresponding to the translational invariance. Thus we need some technical assumptions in order to establish the claim corresponding to Theorem 3.20, which manifests the invariance of the index under deformations of the data. We leave the general formulation to Chapter 5 (cf. Theorem 5.38). Here, we only state the simplest, special, case as the following:

LEMMA 3.28. *Suppose that X is diffeomorphic to the interior of a compact manifold with boundary and a Riemannian metric on X, a pair $(W, [h])$ and an operator D of Dirac type satisfy Conditions* (R) *and* (D). *Then* $\operatorname{ind} D_t$ *is invariant under continuous deformations of the Riemannian metric and h, D only on K_0, whenever Conditions* (R) *and* (D) *are satisfied.*

3. The Excision Theorem and Topological Invariance of the Index

If the end of X is terribly wild, X needs not to be diffeomorphic to the interior of a compact manifold with boundary. We give the way to define the index of the pair $(W, [h])$, even in such a case, by reducing the problem to the case with cylindrical end.

Certainly, we do not, usually, study an operator of Dirac type on such a wild manifold, unless it is the purpose itself. Nevertheless, we could define the index on X, which is diffeomorphic to the interior of a compact manifold with boundary, using various Riemannian metrics at the end of X (although the behavior of which is relatively nice). We would like to show that these definitions coincide. But we need to assume some "technical assumptions" as we mentioned at the end of § 2.4, for the argument, based on deformation by homotopy, following the proof of Theorem 3.20. Thus it is not easy (in analysis) to reduce the problem to continuity of eigenvalues as in Theorem 3.19 by joining two data on X including Riemannian metrics simply by homotopy.

In this section, we explain the method to compare two data, which cannot be joined by homotopy. It solves two problems (how to deal with wild ends and coincidence of definitions of the index) simultaneously. Explicitly, it is called the "Excision Theorem". The Excision Theorem also plays an essential role in the proof of the index theorem.

THEOREM 3.29 (Excision Theorem). *Suppose that X is diffeomorphic to the interior of a compact manifold with boundary and partitioned into a relatively compact open submanifold Y and its complement $X \setminus Y$ by a closed submanifold of codimension* 1. *Then the closure of Y is a compact manifold with boundary and Y is its interior. Let $(W_X, [h_X])$ be a pair on X such that h_X is an isomorphism on $X \setminus Y$. Then the restriction $(W_X|Y, [h_X|Y])$ of $(W_X, [h_X])$ to Y is a pair on Y. We construct the operator $D_{X,t}$ of Dirac type on X using arbitrary Riemannian metrics, etc., satisfying Conditions (R) and (D). On Y, we construct $D_{Y,t'}$ using the Riemannian metric with cylindrical end and the translationally invariant data at the end. Then the following equality holds:*

$$\operatorname{ind} D_{Y,t'} = \operatorname{ind} D_{X,t}.$$

In particular, when Z is the boundary of Y in X and $X = Y \cup Z \times [-1, 0)$, $\operatorname{ind} D_{X,t}$ does not depend on data such as Riemannian metrics whenever Conditions (R) and (D) are satisfied.

We can define the index for general X, which needs not be the interior of a compact manifold with boundary, as follows.

Let $(W, [h])$ be a pair on a manifold X, which may have a terribly wild end. Partitioning X, by a closed manifold of codimension 1, into the relatively compact open submanifold Y_0 and its complement $X \setminus Y_0$, we assume that h is an isomorphism on $X \setminus Y_0$. (We can choose the proper map $f : X \to [-1, \infty)$ so that it takes negative values where h is not isomorphic. Moreover, without loss of generality, we can assume that 0 is a regular value. Then we set $Y_0 = f^{-1}([-1, 0))$.) Then the index $\operatorname{ind}(W|Y_0, [h|Y_0])$ of the restriction makes sense. For Y_1, Y_2, which contain all points where h is not isomorphic in their interior, we can choose Y_0 contained in $Y_1 \cap Y_2$. We define the index on Y_1, Y_2 from the data satisfying Conditions (R) and (D). Applying the Excision Theorem twice, we find that

$$\operatorname{ind}(W|Y_1, [h|Y_1]) = \operatorname{ind}(W|Y_0, [h|Y_0]) = \operatorname{ind}(W|Y_2, [h|Y_2]).$$

Thus we define $\operatorname{ind}(W, [h])$ by this number.

Now we explain the proof of the Excision Theorem. Note that deformations of t are absorbed in deformations of h and that the index is invariant when deforming the data on a compact region (Lemma

3.28). Then we clearly obtain the Excision Theorem from Theorem 3.30, which we explain below.

Suppose that X is diffeomorphic to the interior of a compact manifold with boundary and partitioned by a closed submanifold of codimension 1 into a relatively compact open submanifold Y and its complement $X \setminus Y$. Let $(W_X, [h_X])$ be a pair on X. We assume that h is an isomorphism on $X \setminus Y$.

Suppose that, on X, a Riemannian metric g, a function f, a pair $(W_X, [h_X])$ an operator D_X of Dirac type and a representative h_X of $[h_X]$ satisfy Conditions (R) and (D). If necessary, we deform the Riemannian metric on X, f and D_X, h_X on a compact subset of X so that we may assume the following conditions.

- -2 is a regular value of f and $Y = f^{-1}(-\infty, -2)$.
- $f^{-1}([-2,-1])$ is isometric to $Z_Y \times [-2,-1]$ with some compact Riemannian manifold Z_Y.
- h_X, D_X is translationally invariant on $f^{-1}([-2,-1])$, which is isometric to $Z_Y \times [-2,-1]$.

For each $L > 0$, we replace the part in X, which is isometric to $Z_Y \times [-2,-1]$, by $Z_Y \times [-2, 3L]$ to get a Riemannian manifold X^L. Then a pair $(W^L, [h^L])$ and an operator D^L of Dirac type are naturally defined on X^L, which are translationally invariant on $Z_Y \times [0, 3L]$. In this case, we find that $\operatorname{ind} D^L_t = \operatorname{ind}(W_X, [h_X])$. Meanwhile, replacing the part $f^{-1}([-2,\infty))$ of X by $Z_Y \times [-2, \infty)$, we get a Riemannian manifold, which is denoted by X^∞. X^∞ is diffeomorphic to Y and has the cylindrical end (cf. Figure 1). On this space, the pair $(W^\infty, [h^\infty])$ and D^∞, which is translationally invariant on the cylinder, are naturally defined. Hence we obtain $\operatorname{ind} D^\infty_{X,t} = \operatorname{ind}(W_X|Y, [h_X|Y])$.

For a sufficiently large $t > 0$, we can choose $\bar{\lambda}, \epsilon$ once for all so that the conditions in the previous section are satisfied for all L including ∞.

Under these settings, we have the following

THEOREM 3.30. *Suppose that, for $\mu < \bar{\lambda}$, $(D^\infty_t)^2$ has no eigenfunctions with eigenvalue μ. Then, for a sufficiently large L, the numbers of eigenvalues (for each degree of $W^\bullet$), which are less than or equal to μ, are the same for $(D^\infty_t)^2$ and $(D^L_t)^2$.*

The proof will be given in more general settings in Chapter 5, § 6 (Theorem 5.40).

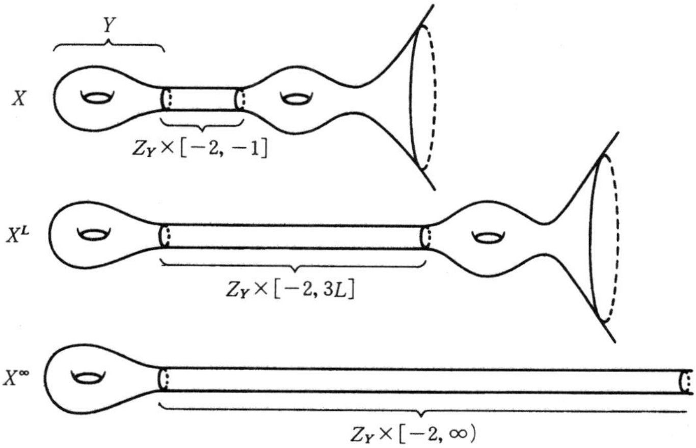

FIGURE 1. X, X^L, X^∞

4. Products of Operators of Dirac Type and Their Indices

The following examples are important with non-cylindrical ends.

EXAMPLE 3.31. Let $(W, [h])$, resp. $(W', [h'])$ be a pair on X, resp. X', and D, resp. D', an operator of Dirac type on W, resp. W', in such a way that Conditions (R) and (D) are satisfied. Denote by $\tilde{W}$ the exterior tensor product $W' \boxtimes W$. Then we can construct a pair $(\tilde{W}, [\tilde{h}])$ on $X' \times X$ and an operator $\tilde{D}$ of Dirac type on $\tilde{W}$ as follows:

$$\begin{aligned} c_{\tilde{W}}(v' \oplus v) &= c_{W'}(v') \otimes \mathrm{id}_W + \epsilon_{W'} \otimes c_W(v), \\ \tilde{h} &= h' \otimes \mathrm{id}_W + \epsilon_{W'} \otimes h, \\ \tilde{D} &= D' \otimes \mathrm{id}_W + \epsilon_{W'} \otimes D. \end{aligned}$$

Firstly, the fact that $\tilde{h}^2 = h'^2 \otimes \mathrm{id}_W + \mathrm{id}_W \otimes h^2$ implies that a point in $X' \times X$, where $\tilde{h}$ is not isomorphic, is a pair of a point in X', where h' is not isomorphic and a point in X, where h is not isomorphic. Hence, the set of such points is surely compact, and we can define a pair $(\tilde{W}, [\tilde{h}])$. It is also easy to see that they satisfy Conditions (R) and (D).

We call the triple $(\tilde{W}, \tilde{D}, \tilde{h})$ the **product** of (W', D', h') and (W, D, h). Even though X and X' have cylindrical ends, $X' \times X$ does not, in general.

We can see the behavior of the index in the construction of the product as follows. By the definition, we find that

$$\tilde{D}_t^2 = {D'}_t^2 \otimes \mathrm{id}_W + \mathrm{id}_{W'} \otimes D_t^2.$$

Hence, any element s in $\operatorname{Ker} \tilde{D}_t^2$ satisfies

$$\begin{aligned} 0 &= \int \langle \tilde{D}_t^2 s, s\rangle \\ &= \int \langle ({D'}_t^2 \otimes \mathrm{id}_W)s, s\rangle + \int \langle (\mathrm{id}_{W'} \otimes D_t^2)s, s\rangle \\ &= \int |(D'_t \otimes \mathrm{id}_W)s|^2 + \int |(\mathrm{id}_{W'} \otimes D)s|^2. \end{aligned}$$

Here, as we explain in Chapter 5, s decays exponentially and the contribution from infinity is 0 when we apply the integration by parts. Therefore we obtain the isomorphism

$$\operatorname{Ker} \tilde{D}_t^2 = \operatorname{Ker} {D'}_t^2 \otimes \operatorname{Ker} D_t^2. \tag{3.4}$$

This is an isomorphism as graded vector spaces.

From this, we immediately obtain the following product formula.

PROPOSITION 3.32. $\operatorname{ind}(\tilde{W}, [\tilde{h}]) = \operatorname{ind}(W', [h']) \operatorname{ind}(W, [h])$.

REMARK 3.33. Technically speaking, $X' \times X$ has the cylindrical end when X is a closed manifold and X' has the cylindrical end. The same is true for a fiber bundle over X with the fiber X'. Taking this into account, it is enough to study Riemannian manifolds with cylindrical ends for the proof of the index theorem for $\mathbb{Z}_2$-graded Clifford modules on closed manifolds.

For a pair $(W, [h])$ on X, h is a section of $H_{\mathbb{Z}_2}(W)$. The above construction of the product presents an example where such an h arises naturally. We explain this point.

Firstly, we assume, for simplicity, that X, X' are closed manifolds and take $h = h' = 0$ in the above construction.

A formally Hermitian endomorphism D' acts on $\Gamma(W')$. Regard $\Gamma(W')$ as an infinite dimensional $\mathbb{Z}_2$-graded vector space equipped with a Hermitian metric. By taking the tensor product with W (from the left), we get a vector bundle of infinite rank on X. Denote it by $\mathcal{W}$, formally. $\mathcal{W}$ naturally carries a structure of $\mathbb{Z}_2$-graded Clifford module as in the case of the tensor product (from the left) by a $\mathbb{Z}_2$-graded vector bundle of finite rank. Moreover, the Dirac operator $\mathcal{D} = \epsilon_{\Gamma(W')} \otimes D$ acts on it.

Write $\tilde{h} = D' \otimes \mathrm{id}_W$, which belongs to $H_{\mathbb{Z}_2}(\mathcal{W})$. Thus, in a formal sense, $(\mathcal{W}, [\tilde{h}])$ is a pair of infinite rank on X and

$$\mathcal{D}_t = \mathcal{D} + t\tilde{h} = \epsilon_{\Gamma(W')} \otimes D + tD' \otimes \mathrm{id}_W$$

is the Dirac operator, the index of which we must study. Note that a section of $\mathcal{W}$ on X is nothing but a section of $W' \boxtimes W$ on $X' \times X$. For $t = 1$, we find that $\mathcal{D}_t = \tilde{D}$.

Therefore, if a $\mathbb{Z}_2$-graded Hermitian vector space and a Hermitian transformation acting on it appear as a finite dimensional approximation of $\Gamma(W')$ and D' acting on it, we can construct a finite dimensional approximation of $\mathcal{W}, \mathcal{D}, \tilde{h}$ to obtain a pair and an operator of Dirac operator in the honest sense.

In short, h is understood to appear as the trace of X' after taking a finite dimensional approximation in the direction of X' of the operator (this is almost the definition of the index), while we regard X as a factor of the product $X' \times X$.

REMARK 3.34. The central issue in the proof of the index theorem in Chapter 6 is to extend the above consideration for a fiber bundle on X with the fiber X', which is more general than the product (with X' the Euclidean space and D' the supersymmetric harmonic oscillator).

5. Supersymmetric Harmonic Oscillator and the de Rham Complex on the Euclidean Space

In the computation of the index, it is the "supersymmetric harmonic oscillator" that plays a role as a unit (or basic object) in some sense. We give three kinds of definitions. In the first, we use the de Rham complex with coefficients in the orientation local system. In the second, we define the n-dimensional supersymmetric harmonic oscillator as the product of n copies of the 1-dimensional supersymmetric harmonic oscillator. In the third, we construct it from the tensor product of two spinor bundles (the last method is only valid in even dimension.)

5.1. 1-dimensional Case. We explain the simplest example of $\mathbb{Z}_2$-graded Clifford modules with compact support and its associated operator of Dirac type. It is the 1-dimensional supersymmetric harmonic oscillator.

Denote by q the coordinate on $\mathbb{R}$. We use the standard Riemannian metric on $\mathbb{R}$. (Then Condition (R) is satisfied.)

Using the irreducible $\mathbb{Z}_2$-graded unitary representation in Example 2.17 in Chapter 2, we define the triple $(W_{\mathbb{R}}, D_{\mathbb{R}}, h_{\mathbb{R}})$ as follows: Define the structure of a Clifford module on the $\mathbb{Z}_2$-graded vector bundle $W_{\mathbb{R}} = \mathbb{R} \times (\mathbb{C} \oplus \mathbb{C})$ on $\mathbb{R}$ as follows.

$$c_{\mathbb{R}}(dq) = \begin{pmatrix} 0 & -1 \\ 1 & 0 \end{pmatrix}.$$

Then

$$D_{\mathbb{R}} = \begin{pmatrix} 0 & -\partial_q \\ \partial_q & 0 \end{pmatrix}$$

is a translationally invariant operator of Dirac type. Here, $\partial_q = d/dq$. Define a Hermitian section $h_{\mathbb{R}}$ of $\operatorname{End} W_{\mathbb{R}}$ by

$$h_{\mathbb{R}}(q) = \begin{pmatrix} 0 & q \\ q & 0 \end{pmatrix},$$

which anti-commutes with the Clifford multiplication. Moreover, h is isomorphic outside of a compact set $\{q = 0\}$, we can consider the pair $(W_{\mathbb{R}}, [h_{\mathbb{R}}])$ and its index. We are going to determine the index $\operatorname{ind}(W_{\mathbb{R}}, [h_{\mathbb{R}}])$.

For $t > 0$, write $D_{\mathbb{R},t} = D_{\mathbb{R}} + th_{\mathbb{R}}$. Then we find the following:

$$D_{\mathbb{R},t} = \begin{pmatrix} 0 & -\partial_q + tq \\ \partial_q + tq & 0 \end{pmatrix},$$

$$D_{\mathbb{R},t}^2 = \begin{pmatrix} -\partial_q^2 - t + t^2q^2 & 0 \\ 0 & -\partial_q^2 + t + t^2q^2 \end{pmatrix}.$$

The operator $-\partial_q^2 + t^2q^2$ is called the **1-dimensional harmonic oscillator**. The eigenvalues of $D_{\mathbb{R},t}^2$ are determined by the following lemma.

LEMMA 3.35 (Eigenvalues of the 1-dimensional harmonic oscillator). *Eigenvalues of $-\partial_q^2 + t^2q^2$ are all positive odd multiples of t:*

$$t,\ 3t,\ 5t,\ \ldots,\ (2k-1)t,\ \ldots,$$

and the multiplicity of each eigenvalue is 1.

PROOF. Since $h^2 \to \infty$ at infinity, it follows from the general argument in Chapter 5, § 4, Corollary 5.28, that there are at most finite eigenvalues less than or equal to any given number $\bar{\lambda}$ counting with multiplicity. (See also Remark 5.42 in Chapter 5.) All eigenvalues are non-negative. Eigenfunctions (and also its derivatives of any order) decay exponentially.

Denote by E_λ the eigenspace corresponding to the eigenvalue λ, which is a finite dimensional vector space.

Define the differential operator a and its formal adjoint a^* by

$$a = \partial_q + tq, \quad a^* = -\partial_q + tq,$$

and note that $aa^* - t = a^*a + t = -\partial_q^2 + t^2q^2$. For an element $\phi \neq 0$ in E_λ, write $\psi = a\phi$. Then we find that

$$\begin{aligned}(-\partial_q^2 + t^2q^2)\psi &= (aa^* - t)a\phi = a\{(a^*a + t) - 2t\}\phi \\ &= a\{(-\partial_q^2 + t^2q^2) - 2t\}\phi = (\lambda - 2t)\psi.\end{aligned}$$

Hence we get a map $a : E_\lambda \to E_{\lambda-2t}$. Meanwhile, since

$$a^*\psi = a^*a\phi = (\lambda - t)\phi,$$

$\psi \neq 0$ as long as $\lambda \neq t$. Thus, whenever $\lambda \neq t$, the map $a : E_\lambda \to E_{\lambda-2t}$ is an injection. Similarly, we have the map $a^* : E_\lambda \to E_{\lambda+2t}$, which is an injection as long as $\lambda \neq -t$.

Combining this and the fact that all eigenvalues are non-negative, we find the following:

(1) Eigenvalues are only multiples of t by a positive odd integer and their multiplicities are the same.
(2) The condition that the eigenvalue of an eigenfunction ϕ is less than or equal to $\lambda = (2k-1)t$ is equivalent to that $a^k\phi = 0$.
(3) Any eigenfunction with the eigenvalue $\lambda = (2k-1)t$ is expressed as $(a^*)^{k-1}\psi$ by some eigenfunction ψ with the eigenvalue t.

These follow from the commutation relation. The general argument in Chapter 5 guarantees the existence of eigenfunctions, hence the multiplicity is at least 1. On the other hand, $a\phi = 0$ is an ordinary differential equation. Thus the multiplicity is at most 1. Therefore the multiplicity for any eigenvalue is 1. □

Using the argument in the proof above, it is not difficult to write down eigenfunctions explicitly. Note that

$$\begin{aligned}a &= \partial_q + tq = e^{-\frac{1}{2}tq^2}\partial_q e^{\frac{1}{2}tq^2}, \\ a^* &= -\partial_q + tq = e^{\frac{1}{2}tq^2}(-\partial_q)e^{-\frac{1}{2}tq^2},\end{aligned}$$

which implies

$$a^k = e^{-\frac{1}{2}tq^2}\partial_q^k e^{\frac{1}{2}tq^2}, \quad (a^*)^k = e^{\frac{1}{2}tq^2}(-\partial_q)^k e^{-\frac{1}{2}tq^2}.$$

The former implies that the space spanned by all eigenfunctions with eigenvalues less than or equal to $(2k-1)t$ coincides with the space of the products of $\exp(-tq^2/2)$ and all polynomials of degree less than or equal to k. In particular, the eigenfunction with the eigenvalue t is a constant multiple of

$$e^{-\frac{1}{2}tq^2}.$$

From the latter, we find that the eigenfunction with the eigenvalue $(2k-1)t$ is more explicitly expressed as a constant multiple of

$$e^{-\frac{1}{2}tq^2}\left(e^{tq^2}(-\partial_q)^{k-1}e^{-tq^2}\right).$$

What is in the parentheses coincides essentially with the so-called Hermitian polynomial. Using this lemma, we can immediately compute the index of $D_{\mathbb{R},t}$.

PROPOSITION 3.36.

$$\mathrm{Ker}(D_{\mathbb{R},t}^2|B_{\mathbb{R}}(W_{\mathbb{R}}^0)) \cong \mathbb{C}, \quad \mathrm{Ker}(D_{\mathbb{R},t}^2|B_{\mathbb{R}}(W_{\mathbb{R}}^1)) = 0,$$
$$\mathrm{ind}(W_{\mathbb{R}}, [h_{\mathbb{R}}]) = 1\,.$$

5.2. Creation and Annihilation Operators. The algebraic structures appearing in $D_{\mathbb{R},t}$ are clarified by introducing mutually Hermitian conjugate operators

$$b = \begin{pmatrix} 0 & 1 \\ 0 & 0 \end{pmatrix}, \quad b^* = \begin{pmatrix} 0 & 0 \\ 1 & 0 \end{pmatrix}$$

as well as a, a^*. Any pair of a or a^* and b or b^* is commutative. Obviously, a commutes with a itself, and b anti-commutes with b (i.e., $b^2 = 0$). As another non-trivial (anti-)commutation relation, we have

$$aa^* - a^*a = 2t, \quad bb^* + b^*b = 1.$$

We call a, b the **annihilation operator** and a^*, b^* the **creation operator**.

Set $Q = ba^*$. Then the formal adjoint of Q is given by $Q^* = ab^*$, which satisfies

$$Q^2 = Q^{*2} = 0, \quad D_{\mathbb{R},t} = Q + Q^*, \quad D_{\mathbb{R},t}^2 = QQ^* + Q^*Q.$$

The $\mathbb{Z}_2$-degree of a, a^* as an operator is 0, and the degree of b, b^* is 1. Hence, the degree of Q, Q^* is 1. The eigenvalues of $D_{\mathbb{R},t}$ can be determined by the formal argument concerning the algebraic structure of the operators, i.e, the algebraic discussion on the harmonic oscillator in Lemma 3.35 as well as the anti-commutation relation for

b, b^*, which governs the matrix part. (It is only the multiplicity of the lowest eigenvalue that is not determined in this way.)

For example, in particular, solutions of $D^2_{\mathbb{R},t} s = 0$ coincide with solutions of $Qs = Q^* s = 0$. This fact follows from the last equality above (by taking its inner product with s). (It is certainly necessary to use the fact that the integration by parts is applicable, which we will show in Chapter 5.)

Then

(1) Applying the injectivity of a^* to $0 = Qs = a^*(bs)$, we obtain $bs = 0$.
(2) Since b^* is injective on $\operatorname{Ker} b$, $0 = Q^* s = b^*(as)$ implies $as = 0$.
(3) Thus $as = bs = 0$. The functions annihilated by a, b are constant multiples of

$$e^{-\frac{1}{2}tq^2}, \quad \begin{pmatrix} 1 \\ 0 \end{pmatrix},$$

respectively, we find that

$$\operatorname{Ker} D^2_{\mathbb{R},t} = \mathbb{C} \begin{pmatrix} e^{-tq^2/2} \\ 0 \end{pmatrix}.$$

The algebraic framework above is equivalent to what is called, in physics, the **1-dimensional supersymmetric harmonic oscillator**.

In fact, we can also understand it has the origin in the de Rham complex on $\mathbb{R}$. Then it is easy to extend it to the n-dimensional case.

5.3. Generalization of the Supersymmetric Harmonic Oscillator. Let v be a vector field on a Riemannian manifold X. We assume that the zero set of v is a compact subset in X.

We would like to define a generalization of the supersymmetric harmonic oscillator on X.

Firstly, we define a $\mathbb{Z}_2$-graded Clifford module W_X over the tangent bundle TX of X as follows.

DEFINITION 3.37.

$$\begin{aligned} W_X^0 &= \bigoplus_{k\equiv 0(2)} \bigwedge^k TX \otimes \mathbb{C}. \\ W_X^1 &= \bigoplus_{k\equiv 1(2)} \bigwedge^k TX \otimes \mathbb{C}. \\ c_{W_X}(v) &= -b_v + b_v^*, \qquad b_v = -v^{\lrcorner}, \quad b_v^* = -v^{\wedge}. \end{aligned}$$

Interpret the definition above using the cotangent bundle T^*X, we would like to find a relation to the de Rham complex.

Using the Riemannian metric, TX and T^*X are isomorphic as vector bundles. Thus, their exterior powers W_X and $\bigwedge^\bullet T^*X$ are isomorphic. However, the construction below is different from this isomorphism. The Riemannian metric induces a volume element $d\,\mathrm{vol}$ on X. The construction of the isomorphism we consider below is based on $d\,\mathrm{vol}$, not the whole information of the Riemannian metric.

Denote by θ_X the orientation local system of X. Write

$$\left|\bigwedge^k T^*X\right| = \bigwedge^k T^*X \otimes \theta_X,$$

then $d\,\mathrm{vol}$, which is induced by the Riemannian metric on X, is a section of $|\bigwedge^n T^*X|$.

By the contraction, not the Riemannian metric, we obtain the following natural isomorphism

$$\bigwedge^n TX \otimes \bigwedge^{n-k} T^*X \overset{\cong}{\to} \bigwedge^k TX.$$

Using this (since $\theta_X^* \cong \theta_X$, naturally), we obtain the isomorphism

$$\left|\bigwedge^n TX\right| \otimes \left|\bigwedge^{n-k} T^*X\right| \overset{\cong}{\to} \bigwedge^k TX.$$

In particular, once $d\,\mathrm{vol}$ is given on X, we obtain the isomorphism $|\bigwedge^{n-k} T^*X| \cong \bigwedge^k TX$. From this, we get the identification

$$\begin{aligned} W_X^0 &= \bigoplus_{k\equiv n(2)} \left|\bigwedge^k T^*X\right| \otimes \mathbb{C}, \\ W_X^1 &= \bigoplus_{k\not\equiv n(2)} \left|\bigwedge^k T^*X\right| \otimes \mathbb{C}. \end{aligned}$$

Note that, in this expression, the Clifford multiplication is given by

$$c_{W_X} = -b_u + b_u^*, \qquad b_u = -u^{\wedge}, \quad b_u^* = -u^{\lrcorner}$$

for an element u in T^*X.

Using the vector field v, the exterior derivative d and its formal adjoint operator d^*, we define $h_{X,v}$ and D_X as follows.

DEFINITION 3.38. Denote by $\tilde{v}$ the differential 1-form corresponding to v via the isomorphism $TX \cong T^*X$ given by the Riemannian metric. We define the triple $(W_X, D_X, h_{X,v})$ by

$$h_{X,v} = b_{\tilde{v}} + b_{\tilde{v}}^*, \quad D_X = d + d^*.$$

Here, we use the expression $W_X = |\bigwedge^\bullet T^*X|$.

D_X is an operator of Dirac type. But it may not satisfy Condition (R), in general. Even though X satisfies Condition (R), $(D_X, h_{X,v})$ may not satisfy Condition (D), in general.

When X is the Euclidean space E, Conditions (R) and (D) are satisfied with an appropriate choice of v.

DEFINITION 3.39. Define the vector field $v : E \to T^*E \cong TE = E \times E$ on E by the diagonal map. We call the triple (W_E, D_E, h_E) in Definition 3.38 with $h_E = h_{E,v}$ the **supersymmetric harmonic oscillator on E** .

The harmonic oscillator on E satisfies Conditions (R) and (D).

If $\tilde{v}$ coincides with the exterior derivative df of some function f, we can also express it in the following way.

LEMMA 3.40. *Suppose that $\tilde{v}=df$. For the triple $(W_X, D_X, h_{X,v})$, write $D_{X,v,t} = D_X + th_{X,v}$, the operator in total. Then it satisfies the following:*

$$D_{X,v,t} = Q + Q^*, \qquad Q = e^{tf} d e^{-tf}, \qquad Q^* = e^{-tf} d^* e^{tf}.$$

PROOF. It must be clear. □

The differential 1-form corresponding to the vector field v, which is used in the definition of the supersymmetric harmonic oscillator, is expressed as $\tilde{v} = d(|q|^2/2)$, where $|q|$ is the distance from the origin. We state the above lemma in this case.

LEMMA 3.41. *Denote the supersymmetric harmonic oscillator on E by $D_{E,t} = D_E + th_E$. Then it satisfies the following:*

$$D_{E,t} = Q + Q^*, \qquad Q = e^{t|q|^2/2} d e^{-t|q|^2/2},$$
$$Q^* = e^{-t|q|^2/2} d^* e^{t|q|^2/2}.$$

Here $|q|^2 : \mathbb{R}^n \to \mathbb{R}$ is the square of the distance from the origin.

In the case that E is 1-dimensional, this definition of the supersymmetric harmonic oscillator coincides with the previous one. This can be seen from the fact that

$$a = e^{-tq^2/2}\partial_q e^{tq^2/2}, \quad a^* = e^{tq^2/2}(-\partial_q)e^{-tq^2/2}, \quad b = -dq^\wedge, \\ b^* = -dq^\lrcorner$$

under the identification

$$W^0 = \left|\bigwedge^1 T^*\mathbb{R}\right| \otimes \mathbb{C} \cong \bigwedge^0 T\mathbb{R} \otimes \mathbb{C},$$
$$W^1 = \left|\bigwedge^0 T^*\mathbb{R}\right| \otimes \mathbb{C} \cong \bigwedge^1 T\mathbb{R} \otimes \mathbb{C}.$$

In particular, we can see that

$$Q = ba^* = e^{\frac{1}{2}tq^2} d e^{-\frac{1}{2}tq^2}, \quad Q^* = ab^* = e^{-\frac{1}{2}tq^2} d^* e^{\frac{1}{2}tq^2}$$

and find that

$$\operatorname{Ker} D_t^2 = \mathbb{C}e^{-\frac{1}{2}t|q|^2}|dq|.$$

REMARK 3.42. The supersymmetric harmonic oscillator is defined over not only $\mathbb{C}$ but also $\mathbb{R}$, already. When we study the index theorem for Clifford modules over the real number field $\mathbb{R}$, we use this fact.

One of the specific features of the supersymmetric harmonic oscillator is that $W_E = \bigwedge^\bullet TE$ carries not only the $\mathbb{Z}_2$-grading, but also the $\mathbb{Z}$-grading coming from the degree of the exterior power.

Neither the Clifford multiplication nor D_E, h_E are homogeneous operations with respect to this $\mathbb{Z}$-grading. They are written as the sum of operators of degree $+1$ and operators of degree -1. However, we have the following.

LEMMA 3.43. *$D_{E,t}^2$ preserves the $\mathbb{Z}$-grading on W_E.*

This is a special case of the following lemma.

LEMMA 3.44. *Let v be a vector field on a Riemann manifold X. Suppose that $\tilde{v} = df$ for some function f on X. For the triple $(W_X, D_X, h_{X,v})$, we define the $\mathbb{Z}$-grading on $W_X = \bigwedge^\bullet TX$ by the degree of the exterior power. Then the operator $D_{X,v,t}^2$ preserves the $\mathbb{Z}$-grading.*

PROOF. The exterior derivative d and its formal adjoint d^* satisfy $d^2 = d^{*2} = 0$. Hence, the square of the conjugation of each of them by the multiplication operator by $\exp \pm tf$ is 0. Note that $D_{X,v,t}^2$

is expressed as the sum of operators of degree $-2, 0, 2$ and its homogeneous part of degree -2, resp. 2 is in such a form. Hence they are 0. □

5.4. Multiplicativity of Supersymmetric Harmonic Oscillator. When E is a direct sum of two subspaces E_0 and E_1, the following multiplicativity easily follows from the definition.

PROPOSITION 3.45. *For* $E = E_0 \oplus E_1$, (W_E, D_E, h_E) *is the product of* $(W_{E_0}, D_{E_0}, h_{E_0})$ *and* $(W_{E_1}, D_{E_1}, h_{E_1})$. *In particular, by* (3.4), $\operatorname{Ker} D^2_{E,t}$ *is isomorphic to the tensor product of* $\operatorname{Ker} D^2_{E_0,t}$ *and* $\operatorname{Ker} D^2_{E_1,t}$.

Here, we consider that Ker consists of smooth solutions exponentially decaying at infinity (as a characterization, for instance). More concretely speaking, $\operatorname{Ker} D^2_{E,t}$ is written as

$$\operatorname{Ker} D^2_{E,t} = \mathbb{C}e^{-t|q|^2/2} d\,\mathrm{vol}.$$

From this, we find that the index of the supersymmetric harmonic oscillator is 1 (cf. Proposition 3.32).

THEOREM 3.46. $\operatorname{ind}(W_E, [h_E]) = 1$.

REMARK 3.47. This fact is related to the de Rham cohomology $H^\bullet_c(E, \theta_E)$ of the complex $(\Omega^\bullet_c(E, \theta_E), d)$ consisting of differential forms with compact support (or, the homology group $H_\bullet(E)$, which is isomorphic to the above by the Poincaré duality). Consider the complex of differential forms of certain exponential decay instead of ones with compact support. The supersymmetric harmonic oscillator is the study of the corresponding de Rham cohomology.

We leave the concrete expression in terms of creation-annihilation operators as an exercise.

EXERCISE 1. Let $(W_{\mathbb{R}}, D_{\mathbb{R}}, h_{\mathbb{R}})$ be the supersymmetric harmonic oscillator on $\mathbb{R}$. Define the $\mathbb{Z}_2$-graded Clifford module $W_{\mathbb{R}^n}$ on $\mathbb{R}^n$ by the exterior tensor product $W_{\mathbb{R}^n} = W_{\mathbb{R}^1} \boxtimes \cdots \boxtimes W_{\mathbb{R}^1}$ and the operators a_k, b_k $(k = 1, 2, \ldots, n)$ on $W_{\mathbb{R}^n}$ by

$$\begin{aligned} a_k &= \overbrace{\epsilon_{W_{\mathbb{R}}} \boxtimes \cdots \boxtimes \epsilon_{W_{\mathbb{R}}}}^{k-1} \boxtimes a \boxtimes \overbrace{\mathrm{id}_{W_{\mathbb{R}}} \boxtimes \cdots \boxtimes \mathrm{id}_{W_{\mathbb{R}}}}^{n-k}, \\ b_k &= \overbrace{\epsilon_{W_{\mathbb{R}}} \boxtimes \cdots \boxtimes \epsilon_{W_{\mathbb{R}}}}^{k-1} \boxtimes b \boxtimes \overbrace{\mathrm{id}_{W_{\mathbb{R}}} \boxtimes \cdots \boxtimes \mathrm{id}_{W_{\mathbb{R}}}}^{n-k}. \end{aligned}$$

We define a_k^*, b_k^* in a similar way.

(1) Find the commutation relation among them.
(2) Show that each fiber of $W_{\mathbb{R}^n}$ is the complexification of a real irreducible representation of the algebra generated by b_k, b_k^* $(k = 1, 2, \ldots, n)$.
(3) Show that the space of sections of $W_{\mathbb{R}^n}$, which are annihilated by the action of all a_k and b_k, is 1-dimensional. Denote the generator by $|0\rangle$, which is called the **vacuum**.
(4) Check that $O(n)$ naturally acts on the algebra generated by b_k, b_k^* $(k = 1, 2, \ldots, n)$. Then, as in Proposition 2.39, show that a projective action of (n) is defined on the real irreducible representation in (2).
(5) Define the action by $O(n)$ on $|0\rangle$ to be trivial and check that the projective action of $O(n)$ is an honest linear representation. Show that it is isomorphic to the complexification of the exterior algebra of $\mathbb{R}^n$.

5.5. Relation with Spinor Bundles. We express the supersymmetric harmonic oscillator in terms of spinor bundles. The details are left to readers.

Let E be an n-dimensional vector space equipped with a Euclidean metric. Suppose that R is a unitary representation of the Clifford algebra $Cl(E \oplus E)$ over $E \oplus E$. Then we define the structure of a Clifford module on the vector bundle $W_R = E \times R$ on E, the operator D_{W_R} of Dirac type on W_R and the Hermitian endomorphism h_{W_R}, which anti-commutes with the Clifford multiplication as follows.

$$c_{W_R}(v) = c_R(0 \oplus v), \tag{3.5}$$

$$D_{W_R} s = \sum_{k=1}^{n} (c_{W_R}(e_k))(\partial_{e_k} s), \tag{3.6}$$

$$h_{W_R}(q) = \sqrt{-1} c_R(v(q) \oplus 0)\,. \tag{3.7}$$

Here, $\{e_k\}$ is the orthonormal frame determined by orthogonal coordinates on E and v is the one in Definition 3.39.

Since the support of h_{W_R} is only the origin, we obtain the pair $(W_R, [h_{W_R}])$.

Since $E \oplus E$ is even dimensional, Theorem 2.11 implies that R is isomorphic to the direct sum of copies of the spinor representation $\Delta_{E \oplus E}$ of $Cl(E \oplus E)$.

LEMMA 3.48. *For $R = \Delta_{E \oplus E}$, the triple (W_R, D_R, h_R) is isomorphic to the supersymmetric harmonic oscillator.*

In the case that the dimension n of E is even, we can rephrase the expression in Lemma 3.48 as follows.

Suppose that $n = 2m$. The spinor representation $\Delta = \Delta^0 \oplus \Delta^1$ of $Spin(2m)$ gives rise to the spinor bundle $S_{\mathbb{R}^{2m}} = \mathbb{R}^{2m} \times \Delta$ on $\mathbb{R}^{2m}$.

On the one hand, we regard it just as a $\mathbb{Z}_2$-graded vector bundle and define the endomorphism using the Clifford multiplication. On the other hand, we regard it as a $\mathbb{Z}_2$-graded Clifford module. Tensoring them, we get a $\mathbb{Z}_2 \times \mathbb{Z}_2$-graded Clifford module. This is the desired expression.

More concretely, we can state the result as follows.

PROPOSITION 3.49. *$W_{\mathbb{R}^{2m}}$ is naturally isomorphic to $S_{\mathbb{R}^{2m}} \otimes S_{\mathbb{R}^{2m}}$. Denote by $(q_1, q_2, \dots, q_{2m})$ the coordinate on $\mathbb{R}^{2m}$ and by $e_1, e_2, \dots, e_{2m}$ the corresponding orthonormal frame. Then we obtain the following expression:*

$$\begin{aligned} c_{W_{\mathbb{R}^{2m}}}(e_k) &= \epsilon_{S_{\mathbb{R}^{2m}}} \otimes c_{S_{\mathbb{R}^{2m}}}(e_k), \\ D_{\mathbb{R}^{2m}} &= \epsilon_{S_{\mathbb{R}^{2m}}} \otimes \sum_{k=1}^{2m} c_{S_{\mathbb{R}^{2m}}}(e_k)\partial_{q_k}, \\ h_{\mathbb{R}^{2m}} &= \sqrt{-1} \sum_{k=1}^{2m} q_k c_{S_{\mathbb{R}^{2m}}}(e_k) \otimes \mathrm{id}_{S_{\mathbb{R}^{2m}}} . \end{aligned}$$

To show this, it is enough to check some commutation relations while taking the Exercise in the previous section into account.

SUMMARY.

3.1 For an open manifold with suitable behavior at infinity and an operator of Dirac type on it, we can directly define the index by introducing appropriate function spaces.

3.2 Using the localization of eigenfunctions of the operator of Laplace type, we can establish the Excision Theorem for the index of the operators of Dirac type. The Excision Theorem appears as one of the steps, when we prove the index theorem later.

3.3 By the Excision Theorem, we can define the index for $\mathbb{Z}_2$-graded Clifford modules with compact support on general open manifolds.

3.4 The product formula holds for the index. It is one of the steps in the proof of the index theorem given below to show the fact for vector bundles corresponding to the product formula.

3.5 By "twisting" the de Rham complex on the Euclidean space with coefficient in the orientation local system by the exponential function of the square of the distance from the origin, we define the supersymmetric harmonic oscillator. Its index is 1 and it plays the role of the "unit" among operators of Dirac type.

CHAPTER 4

Examples of the Localization of the Index

In this chapter, we discuss the Poincaré-Hopf theorem and the Riemann-Roch theorem on Riemann surfaces as examples of the index, which can be computed by the Excision Theorem.

As we explained in Chapter 1, there are some variations in the notion of the index. By the Excision Theorem, or by directly applying the localization, which is just one step toward the Excision Theorem, we study the modulo 2-index for spin structures and the Lefschetz formula under group actions on Riemann surfaces.

When the Clifford module carries a $\mathbb{Z}$-grading and is related to some complex, we sometimes obtain the information on the dimension of the cohomology, which is finer than the index, based on the localization argument. As such an example, we present Witten's approach to the Morse inequality.

1. The Poincaré-Hopf Theorem and the Morse Inequality

We study the index of the operator of Dirac type associated with the de Rham complex[1].

1.1. The Poincaré-Hopf Theorem. Let v be a vector field on an n-dimensional Riemannian manifold X.

Consider the triple $(W_X, D_X, h_{X,v})$, which was defined, in Definition 3.38 in Chapter 3, as an extension of the supersymmetric harmonic oscillator.

In this subsection, we study the index of the pair $(W_X, [h_{X,v}])$.

The support of $h_{X,v}$ coincides with the zeros of v. By the Excision Theorem, the index is determined only by the data in a neighborhood of the zeros of v.

[1]The localization argument, which we discuss in this section, was initiated by Witten and has broad influence in direct and indirect ways. This book is also influenced by this argument.

We consider the case that the behavior around the zeros is the simplest. We assume that there are only finitely many zeros of v and that v is transversal to the zero section at each zero. For each zero x of v, denote also by $(dv)_x$ the composition of the derivative $(dv)_x : (TX)_x \to (T(TX))_{(x,0)}$ of $v : X \to TX$ at x and the projection $(T(TX))_{(x,0)} \to (TX)_x$. Then, by the assumption, we obtain an automorphism

$$[(dv)_x : \ (TX)_x \to (TX)_x]$$

of the n-dimensional vector space.

The set of automorphisms of the n-dimensional vector space consists of two connected components, which are distinguished by the sign of the determinant.

THEOREM 4.1. *If there are only finitely many zeros of v and all of them are transversal, then the index of $(W_X, [h_{X,v}])$ is equal to the difference between the number of zeros where $\det(dv)_x$ is positive and the number of zeros where $\det(dv)_x$ is negative:*

$$\begin{aligned} \operatorname{ind}(W_X, [h_{X,v}]) &= \#\{x \mid v(x) = 0, \ \det(dv)_x > 0\} \\ &\quad - \#\{x \mid v(x) = 0, \ \det(dv)_x < 0\}. \end{aligned}$$

Before giving the proof, we state the case of the closed manifold in the following corollary.

COROLLARY 4.2 (Poincaré-Hopf Theorem). *Let X be an n-dimensional closed manifold and v a vector field on X with only finitely many zeros, all of which are transversal. Then the following holds:*

$$\begin{aligned} \sum_k (-1)^{n-k} \dim H^k(X, \theta_X) &= \#\{x \mid v(x) = 0, \ \det(dv)_x > 0\} \\ &\quad - \#\{x \mid v(x) = 0, \ \det(dv)_x < 0\}. \end{aligned}$$

PROOF. By Theorem 4.1, the right hand side is the index of $D_{X,v,t}$ with $t > 0$. If X is a closed manifold, we can also use the case that $t = 0$ in order to compute the index. By Theorem 5.45, which we will prove in Chapter 5, we find that

$$[\operatorname{Ker} D_X \cong H^{n-\bullet}(X, \theta_X)]$$

as $\mathbb{Z}_2$-graded vector spaces. Hence the index of D_X is equal to the left hand side. □

REMARK 4.3.

(1) For any rank r local system θ on X, we can apply the argument in this subsection to the de Rham complex with coefficients in θ in almost the same way. For example, if X is a closed manifold and v is a vector field on X only with finitely many zeros, all of which are transversal, then the following holds:

$$\begin{aligned}\sum_k (-1)^{n-k} \dim H^k(X,\theta) \\ = r(\#\{x \mid v(x)=0,\ \det(dv)_x > 0\} \\ -\#\{x \mid v(x)=0,\ \det(dv)_x < 0\}).\end{aligned}$$

(2) If X is a closed manifold, $H^k(X, \theta_X)$ is isomorphic to the $n-k$-th homology group $H_{n-k}(X, \mathbb{R})$ with coefficients in $\mathbb{R}$ by the Poincaré duality. Define the **Euler number** $\chi(X)$ of X by

$$\chi(X) = \sum_k (-1)^k \dim H_k(X, \mathbb{R}).$$

Then the Poincaré-Hopf theorem can be stated as follows:

$$\begin{aligned}\chi(X) = \#\{x \mid v(x)=0,\ \det(dv)_x > 0\} \\ - \#\{x \mid v(x)=0,\ \det(dv)_x < 0\}.\end{aligned}$$

PROOF OF THEOREM 4.1. By the Excision Theorem, the index is determined by the data around zeros of v. Hence, the problem reduces to the case that X is the Euclidean space and the origin is the unique transversal zero of v.

By continuous deformations of Riemannian metrics on X and v, we may assume that X is isometric to the n-dimensional Euclidean space $\mathbb{R}^n$ and that v is linear. Namely, we assume that v is written as

$$v(q^1, q^2, \ldots, q^n) = \sum_{ij} a_{ij} q^i \partial_{q^j}$$

with some $A = (a_{ij})$ in $GL_n(\mathbb{R})$. Here, $(q^1, q^2, \ldots, q^n)$ are the coordinates on $\mathbb{R}^n$. (Note that we adopt the upper indices. Do not confuse with the degree of the power.)

There are two connected components of $GL_n(\mathbb{R})$, and they are distinguished by the sign of the determinant. Thus, it is enough to show that there are at least one A, with positive determinant, resp. negative determinant, which satisfies the conclusion, since the theorem reduces to such cases. Such A's are given in the examples below. □

The remaining task in the proof above is to present examples of v, which is linear on the Euclidean space, and the matrix A, which determines v, has positive, resp. negative, determinant. The examples below are such A's.

Let X be an n-dimensional Euclidean space $\mathbb{R}^n$ and $n^+, n^- \geq 0$ be the integers such that $n = n^+ + n^-$ is non-negative. Denote the n coordinate functions on $\mathbb{R}^n$ by $q_+^1, q_+^2, \ldots, q_+^{n^+}$ and $q_-^1, q_-^2, \ldots, q_-^{n^-}$.

Define the vector field $v(n^+, n^-)$, the zero of which is only at the origin, by

$$v(n^+, n^-) = -\sum_{i=1}^{n^-} q_-^i \partial_{q_-^i} + \sum_{j=1}^{n^+} q_+^j \partial_{q_+^j}.$$

Then, for $x_0 = 0$, we find that

$$(dv(n^+, n^-))_{x_0} = \begin{pmatrix} -I_{n^-} & O \\ O & I_{n^+} \end{pmatrix}$$

and the sign of its determinant is determined by the parity of n^-.

The differential 1-form corresponding to $v(n^+, n^-)$ is

$$\tilde{v}(n^+, n^-) = -\sum_{i=1}^{n^-} q_-^i dq_-^i + \sum_{j=1}^{n^+} q_+^j dq_+^j$$

and also written as $\tilde{v}(n^+, n^-) = df_{n^+,n^-}$ with the function

$$f_{n^+,n^-} = -\frac{1}{2}\sum_{i=1}^{n^-} (q_-^i)^2 + \frac{1}{2}\sum_{j=1}^{n^+} (q_+^j)^2.$$

Let $(W_{\mathbb{R}}, D_{\mathbb{R}}, h_{\mathbb{R}})$ be the supersymmetric harmonic oscillator on $\mathbb{R}$. Using the expression in coordinates, we find the following.

LEMMA 4.4. *The triple* $(W_{\mathbb{R}^n}, D_{\mathbb{R}^n}, h_{\mathbb{R}^n, v(n^+, n^-)})$ *is isomorphic to the product of* n^+*-copies of* $(W_{\mathbb{R}}, D_{\mathbb{R}}, h_{\mathbb{R}})$ *and* n^-*-copies of* $(W_{\mathbb{R}}, D_{\mathbb{R}}, -h_{\mathbb{R}})$.

Here, $(W_{\mathbb{R}}, D_{\mathbb{R}}, -h_{\mathbb{R}})$ is seen, in the expression in coordinates, to be isomorphic to the 1-dimensional supersymmetric harmonic oscillator with the grading exchanged between the degree 0 part and the degree 1 part. (More precisely, in the intrinsic definition independent of coordinates, they are not isomorphic. But the only difference is the action of $O(1) = \{\pm 1\}$.) For $t > 0$, write $D_{\mathbb{R},-t} = D_{\mathbb{R}} - th_{\mathbb{R}}$. Then the degree 0 part of $\mathrm{Ker}(D_{\mathbb{R},-t})^2$ is 0-dimensional and the degree 1 part is 1-dimensional. From the product formula in Chapter 3, § 4, we immediately obtain the following proposition and its corollary.

PROPOSITION 4.5. *For $t > 0$, the vector space*

$$\mathrm{Ker}\, D^2_{\mathbb{R}^n, v(n^+,n^-),t} \cong (\mathrm{Ker}\, D^2_{\mathbb{R},t})^{\otimes n^+} \otimes (\mathrm{Ker}\, D^2_{\mathbb{R},-t})^{\otimes n^-}$$

is 1*-dimensional as a $\mathbb{Z}_2$-graded vector space. The $\mathbb{Z}_2$-degree of the* 1*-dimensional part is* $n^- \mod 2$.

COROLLARY 4.6. $\mathrm{ind}(\mathbb{R}^n, [h_{\mathbb{R}^n, v(n^+,n^-)}]) = (-1)^{n^-}$.

This corollary provides the examples, which are necessary in the proof of Theorem 4.1.

1.2. Morse Inequality. We return to general X and v. We assumed that v has only finitely many zeros, all of which are transversal.

In this subsection, we consider the case that the differential 1-form, corresponding to v via the Riemannian metric, is written as *df* for some function f. Define the $\mathbb{Z}$-grading on $W_X = \bigwedge^\bullet TX$ by the degree of the exterior power. Then $D^2_{X,v,t}$ preserves the $\mathbb{Z}$-grading by Lemma 3.44 in Chapter 3. In this section, we investigate finer information than the index by the consideration on the $\mathbb{Z}$-grading. We study the localization to a neighborhood of zeros of *df* in detail.

We assume that the Riemannian metric on X and the operator D_X, $h_{X,v}$ satisfy Conditions (R) and (D).

1.2.1. *Morse function and Morse's lemma.* Firstly, we prepare terminologies, which are necessary in the later argument.

When *df* is 0 at a point x, we call x a **critical point** of f. Denote by $(x, 0)$ the origin 0 in the cotangent space $(T^*X)_x$ of X at x. Then we have a natural decomposition

$$T(T^*X)_{(x,0)} \cong TX_x \oplus (T^*X)_x.$$

For a critical point x of f, the derivative of $df : X \to T^*X$ at x is a linear map from $(TX)_x$ to $T(T^*X)_{(x,0)}$. Its composition with the projection to $(T^*X)_x$ is regarded as a bilinear map on $(TX)_x$. By expressing it in the local coordinates, we can see that it is a symmetric bilinear form (by the commutativity of partial differentiation $\partial_{q^1}\partial_{q^2} = \partial_{q^2}\partial_{q^1}$). This symmetric bilinear form is called the **Hessian matrix**. We denote it by Hess^f_x. Unless any confusion arises, we also denote by Hess^f_x the quadratic form $u \mapsto \mathrm{Hess}^f_x(u,u)/2$ on TX determined by Hess^f_x. By the Riemannian metric, we identify TX and T^*X and use the vector field v corresponding to $\tilde{v} = df$. Then the Hessian matrix at the critical point x coincides with $(dv)_x$.

DEFINITION 4.7.

(1) If the Hessian matrix is non-degenerate at all critical points of f, we call f a **Morse function**.
(2) Let x be a non-degenerate critical point of f. Denote by n_x^-, resp. n_x^+, the dimension of a maximal subspace, on which the quadratic form Hess_x^f is negative definite, resp. positive definite. We call n_x^- the **Morse index** of f at x.

Pick an appropriate basis of $(TX)_x$ to get an isomorphism $(TX)_x \cong \mathbb{R}^n$. Then Hess_x^f is equal to $f_{n_x^+, n_x^-}$. Here, n is the dimension of X.

For a critical point x_0, where the Hessian matrix is non-degenerate, it is known that f is identified with $\mathrm{Hess}_{x_0}^f$ up to a constant term using an appropriate coordinate on a neighborhood of x_0.

LEMMA 4.8 (Morse's Lemma). *For a non-degenerate critical point x of f, there is a diffeomorphism ϕ from a neighborhood of x to a neighborhood of 0 in $(TX)_x$ such that the following condition holds:*

$$\phi(x) = 0, \quad (d\phi)_x = \mathrm{id}_{(TX)_x}, \quad f = f(x) + \phi^* \mathrm{Hess}_x^f .$$

We skip the proof of this lemma.

REMARK 4.9. When f has non-degenerate critical points, we can easily find a function without changing the position of critical points and the Hessian matrices at them, which enjoy the property in Morse's lemma. Although we do not prove the lemma, this fact is enough for our purpose in this subsection.

1.2.2. *The Witten complex and the Morse inequality.* For simplicity, we assume that X is a closed manifold. In order to make the $\mathbb{Z}$-grading more tractable, we introduce the following notation:

$$\begin{aligned} \Omega_k(X) &:= \Omega^{n-k}(X, \theta_X), \\ H_k(X) &:= H^{n-k}(X, \theta_X). \end{aligned}$$

(We adopted the homological notation by taking the Poincaré duality into account, but it is just a convention here.) $H_\bullet(X)$ is the homology of the complex $(\Omega_\bullet(X), d)$.

For a Morse function f on X, we can construct a complex, the homology of which is $H_\bullet(X)$, as a finite dimensional approximation of the complex $(\Omega_\bullet(X), d)$. We call it the **Witten complex**.

The construction consists of steps (1), (2) and (3).

(1) "Twist" the de Rham complex by the conjugation of e^{tf}.

For a function f on X and a positive real number $t > 0$, we set $Q = e^{tf} d e^{-tf}$, which satisfies $Q^2 = 0$, and the homology of the complex $(\Omega_\bullet(X), Q)$ is isomorphic to the homology $H_\bullet(X)$ of the complex $(\Omega_\bullet(X), d)$ with the $\mathbb{Z}$-grading taken into account. The isomorphism is given by the multiplication by e^{tf}:

$$H_\bullet(X) = H_\bullet(\Omega_\bullet(X), d) \xrightarrow{\cong} H_\bullet(\Omega_\bullet(X), Q), \quad [s] \mapsto [e^{tf} s].$$

(2) Express the homology of the twisted complex as Ker of the operator of Dirac type.

The principal symbols of two first order differential operators d, Q are the same. Hence $Q + Q^*$ is also an operator of Dirac type, as is $d + d^*$. By Lemma 3.40 in Chapter 3, $Q + Q^*$ is equal to $D_{X,v,t}$ with $\tilde{v} = df$. It follows from Theorem 5.45 (and also Remark 5.46), which are proved in Chapter 5, that we obtain the isomorphism

$$H_\bullet(\Omega_\bullet(X), Q) \cong \operatorname{Ker}(D_{X,v,t})^2$$

as $\mathbb{Z}$-graded vector spaces.

(3) Instead of Ker, we consider the eigenfunctions with small eigenvalues and define a finite dimensional approximation.

Fix an arbitrary number $\bar{\lambda} > 0$. Denote by $E_{\leq \bar{\lambda}, \bullet}(D^2_{X,v,t})$ the vector space spanned by all eigenfunctions with the eigenvalues of $D^2_{X,v,t}$, which is less than or equal to $\bar{\lambda}$. Here $\bullet$ stands for the $\mathbb{Z}$-grading. By Theorem 5.43, which will be proved in Chapter 5, this is finite dimensional. Since $D^2_{X,v,t}$ commutes with Q, we obtain a finite dimensional complex

$$(E_{\leq \bar{\lambda}, \bullet}(D^2_{X,v,t}), Q)$$

by restricting Q. It is easily to see that its homology is isomorphic to $\operatorname{Ker}(D_{X,v,t})^2$ for any $\bar{\lambda} > 0$.

PROPOSITION 4.10. *For a Morse function f on a closed manifold X, there exists a complex, which is obtained as a finite dimensional approximation, such that the dimension of the degree k part with respect to the $\mathbb{Z}$-grading is equal to the number of critical points of Morse index k. More precisely, for a fixed $t > 0$, we pick a sufficiently small positive number $\mu > 0$ (in fact, it is sufficient that $0 < \mu < t$), so that the following holds:*

$$\dim E_{\leq c^2\mu, \bullet}(D^2_{X,v,c^2 t}) = \#\{x \in X \mid (df)_x = 0,\ n_x^- = k\}.$$

In the next subsection, we will give a proof of this proposition. Before that, we note the following corollary.

COROLLARY 4.11 (Morse inequality). *Let X be a closed manifold and f a Morse function on X. There is a polynomial $R(z)$ of one variable with non-negative coefficients such that the following holds:*

$$\sum_k z^k \#\{x \in X \mid (df)_x = 0,\ n_x^- = k\} - \sum_k z^k \dim H_k(X)$$
$$= (1+z)R(z).$$

In particular, the following inequality holds for each k:

$$\dim H_k(X) \le \#\{x \in X \mid (df)_x = 0,\ n_x^- = k\}.$$

PROOF. In general, when $H_\bullet$ is the homology of a finite dimensional complex $(C_\bullet, \partial)$, we write $B_\bullet = \operatorname{Im} \partial$ to obtain $\dim C_k - \dim H_k = \dim B_k + \dim B_{k-1}$. Using $R(z) = \sum_k z^k \dim B_k$, we present it in the form of a generating function, which is the desired formula. □

Substitute $z = -1$ to reproduce the Poincaré-Hopf theorem for $\tilde{v} = df$.

1.2.3. *Localization Argument.* The point in the proof of Proposition 4.10 is that, when t is sufficiently large, the support of eigenfunctions of $(D_{X,v,t})^2$ with eigenvalues less than or equal to $\bar{\lambda}$ is localized to a neighborhood of zeros of $v = df$.

First of all, we describe the behavior in the neighborhood of critical points.

By choosing a Riemannian metric g_X on X appropriately, we may assume the following conditions by Morse's Lemma (or Remark 4.9).

(1) There exists an isometric embedding $q_x : U_x \to \mathbb{R}^n$ from a neighborhood U_x of each critical point x of f to the Euclidean space $\mathbb{R}^n$.
(2) The restriction of f to U_x coincides with the pull-back of $f(x) + f_{n_x^+, n_x^-}$ by q_x.

Here the restriction of $D_{X,v,t}$ to U_x and the restriction of $D_{\mathbb{R}^n, v(n_x^+, n_x^-), t}$ to the neighborhood $q_x(U_x)$ of the origin are isomorphic. The function f_{n^+,n^-} on the n-dimensional Euclidean space $\mathbb{R}^n$ plays a fundamental role in the study of Morse functions. The argument concerning $D_{\mathbb{R}^n, v(n^+,n^-), t}$ in the previous section is also valid, when we replace the $\mathbb{Z}_2$-grading by the $\mathbb{Z}$-grading.

LEMMA 4.12.

(1) *The triple* $(W_{\mathbb{R}^n}, D_{\mathbb{R}^n}, h_{\mathbb{R}^n,v(n^+,n^-)})$ *is isomorphic to the product of* n^+*-copies of* $(W_{\mathbb{R}}, D_{\mathbb{R}}, h_{\mathbb{R}})$ *and* n^-*-copies of* $(W_{\mathbb{R}}, D_{\mathbb{R}}, -h_{\mathbb{R}})$ *with the* $\mathbb{Z}$*-grading structures.*

(2) *For* $t > 0$,

$$\operatorname{Ker} D^2_{\mathbb{R}^n,v(n^+,n^-),t} \cong (\operatorname{Ker} D^2_{\mathbb{R},t})^{\otimes n^+} \otimes (\operatorname{Ker} D^2_{\mathbb{R},-t})^{\otimes n^-}$$

is 1*-dimensional as a* $\mathbb{Z}$*-graded vector space.*

Next, we consider the behavior when t gets large.

For the differential operator $D_{\mathbb{R}^n,v(n^+,n^-),t}$ on $\mathbb{R}^n$, the dilation of $\mathbb{R}^n$ by multiplying a constant has the same effect as multiplying a constant to t.

LEMMA 4.13. *For a positive number* $c > 0$*, the change of variables* $q \mapsto q'$, $q = cq'$ *on* $\mathbb{R}^n$*, transforms* $D_{\mathbb{R}^n,v(n^+,n^-),t}$ *into* $c^{-1}D_{\mathbb{R}^n,v(n^+,n^-),c^2t}$.

Thus, for any $c > 1$, the restriction of $c^{-1}D_{X,v,c^2t}$ to a neighborhood U_x of the critical point x is isomorphic to the restriction of $D_{\mathbb{R}^n,v(n_x^+,n_x^-),t}$ to the neighborhood $cq_x(U_x)$ of the origin.

Firstly, we present an intuitive explanation.

- For each c, rescale the Riemannian metric by the multiplication by c. Imagine that a person looks, in this scale, around her/himself at the critical point.
- For her/him, a tiny neighborhood around her/his standing point is isometric to the flat Euclidean space and equipped with an operator, which is isomorphic to the standard operator $D_{\mathbb{R}^n,v(n^+,n^-),t}$ on the Euclidean space. However, X, as well as the structure on it, is globally curved far away.
- As c gets larger, such a neighborhood around her/him also gets larger and the flat part extends. If c tends to infinity, it becomes flat as far as one can see, and it is exhausted by the standard operator $D_{\mathbb{R}^n,v(n^+,n^-),t}$ on the Euclidean space.

In order to formulate the intuitive argument above, we write

$$(c^{-1}D_{X,v,c^2t})^2 = (c^{-1}D_X)^2 + V_{t,c}, \quad V_{t,c} = tV_0 + c^2t|v|^2.$$

Then, precisely speaking, the facts, which are necessary, are the following.

- $c^{-1}D_X$ is the operator of Dirac type with respect to the rescaled Riemannian metric by multiplying c (when the Riemannian metric is rescaled by c, the Clifford multiplication is multiplied by c^{-2}).
- $V_{t,c}$ can be arbitrarily large outside of the neighborhoods U_x of critical points x, if c is chosen large enough.
- The neighborhood $cq_x(U_x)$ of the origin in $\mathbb{R}^n$ can contain a ball, centered at the origin, of an arbitrarily large radius, if c is chosen large enough.

In the situation above, Theorem 5.40 in Chapter 5 (its extension to the case of the Euclidean space-like end, not of the cylindrical end; cf. Remark 5.42 in Chapter 5) guarantees the following properties.

PROPOSITION 4.14. *Let $\mu > 0$ be a positive number, which does not coincide with any eigenvalue of $D^2_{\mathbb{R}^n,v(n_x^+,n_x^-),t}$ at any critical point x. Then the following equality holds for a sufficiently large c:*

$$\dim E_{\leq\mu,\bullet}(c^{-2}D^2_{X,v,c^2t}) = \sum_{(df)_x=0} \dim E_{\leq\mu,\bullet}(D^2_{\mathbb{R}^n,v(n_x^+,n_x^-),t}).$$

Proposition 4.10 is deduced from this proposition as follows.

PROOF OF PROPOSITION 4.10. By the second half of Lemma 4.12, we find that, if $\mu > 0$ is chosen sufficiently close to 0,

$$\begin{aligned}\dim E_{\leq\mu,k}(D^2_{\mathbb{R}^n,v(n_x^+,n_x^-),t}) &= \dim\left(\operatorname{Ker} D^2_{\mathbb{R}^n,v(n_x^+,n_x^-),t}\right)_k \\ &= \begin{cases} 1, & (k = n^-) \\ 0, & (k \neq n^-). \end{cases}\end{aligned}$$

Hence, for each k, we have

$$\sum_{(df)_x=0} \dim E_{\leq\mu,k}(D^2_{\mathbb{R}^n,v(n_x^+,n_x^-),t}) = \#\{x \mid (df)_x = 0,\ n_x^- = k\}.$$

On the other hand, note, by the definition, that $E_{\leq\mu,\bullet}(c^{-2}D^2_{X,v,c^2t}) = E_{\leq c^2\mu,\bullet}(D^2_{X,v,c^2t})$ to ensure Proposition 4.10. □

1.3. Deformation of the Supersymmetric Harmonic Oscillator. What we explain in this section are genuinely technical issues.

When we use the supersymmetric harmonic oscillator in the proof of the index theorem in Chapter 5, we need to choose one of following two options.

(1) Study the class of Riemannian metrics and Dirac operators larger than those satisfying Conditions (R) and (D).
(2) Use operators which share the common properties with the supersymmetric harmonic oscillator but which enjoy better properties at the end as a substitute for the supersymmetric harmonic oscillator.

For a systematic description, it is preferable to follow the first strategy to formulate the problem in a larger class than those satisfying Conditions (R) and (D), and it is, in fact, not so difficult. However, the description gets unnecessarily complicated, and the line of the proof may become less transparent. Thus, in order to avoid such a problem, we adopt the second strategy. (For readers who are interested in analytical details, it is a good exercise to find necessary conditions in the first strategy and establish the argument in Chapter 5 under these conditions with reference to Chapter 6 under these conditions; cf. Chapter 3, Remark 3.27.)

We summarize the properties of the supersymmetric harmonic oscillator $(W_{\mathbb{R}^m}, D_{\mathbb{R}^m}, h_{R^m})$, which will be used in the proof of the index theorem in Chapter 6.

LEMMA 4.15.

- *$O(m)$ acts on $W_{\mathbb{R}^m}$ and $D_{\mathbb{R}^m}$, $h_{\mathbb{R}^m}$ are $O(m)$-equivariant.*
- *Fix a sufficiently large $t > 0$. If we consider* Ker *of $D^2_{\mathbb{R}^m,t}$ in the class of sections, which converges to 0 at infinity in a suitable sense, there exists an $O(m)$-invariant section $\bar{b}_t \neq 0$ of $(W_{\mathbb{R}^m})^0$ such that*

$$\operatorname{Ker}(D^2_{\mathbb{R}^m,t} \mid (W_{\mathbb{R}^m})^0) = \mathbb{C}\bar{b}_t, \quad \operatorname{Ker}(D^2_{\mathbb{R}^m,t} \mid (W_{\mathbb{R}^m})^1) = 0.$$

In sum, intuitively speaking, the index of the supersymmetric harmonic oscillator is not just 1 as a number, but

- 1 as the index taking the $O(m)$-action into account.
- not just the difference of dimension, but Ker itself, is "1".

These two properties are enjoyed by operators which are not necessarily the original supersymmetric harmonic oscillator. The supersymmetric harmonic oscillator is defined on $\mathbb{R}^m$ equipped with the standard Riemannian metric. For $m > 1$, the end of $\mathbb{R}^m$ is not cylindrical. But it enjoys the above two properties and there exists a substitute, which has cylindrical end with the translational symmetry.

In the rest of this subsection, we explain the construction of the "substitute".

Choose a Riemannian metric $g = g_{\mathbb{R}^m}$ on $\mathbb{R}^m$ and a Morse function $f = f_{\mathbb{R}^m}$ on it as explained just before Lemma 3.24 in Chapter 3.

Denote by $\mathbb{R}^m_g$ the Riemannian space $\mathbb{R}^m$ with the Riemannian metric g in the expression. $\mathbb{R}^m_g$ has a cylindrical end and $O(m)$ acts isometrically on it with the origin fixed. Using the vector field v with $df = \tilde{v}$, we twist the de Rham complex with the orientation local system, as explained in Definition 3.38 in Chapter 3, to get an extension of the supersymmetric harmonic oscillator, which we denote by $(W_{\mathbb{R}^m_g}, D_{\mathbb{R}^m_g}, h_{\mathbb{R}^m_g, v})$.

LEMMA 4.16. *$\mathbb{R}^m_g$ and $(W_{\mathbb{R}^m_g}, D_{\mathbb{R}^m_g}, h_{\mathbb{R}^m_g, v})$ defined on it satisfy the following* (1) *and* (2) *by replacing g by $c^2 g$, if necessary.*

(1) *They are invariant under the translation on the cylinder and satisfy Conditions* (R) *and* (D).
(2) *They enjoy the two properties in Lemma* 4.15.

PROOF. The first assertion is clear by the construction. The origin is the only critical point of f, and its Morse index is 0. Hence the second assertion follows from Proposition 4.14. □

DEFINITION 4.17. We temporarily call the substitute with the translational invariance on the cylinder constructed in the above lemma the **supersymmetric harmonic oscillator with a cylindrical end**.

Unless any confusion occurs, we use the same notation $(W_{\mathbb{R}^m}, D_{\mathbb{R}^m}, h_{\mathbb{R}^m})$ for the supersymmetric harmonic oscillator with a cylindrical end.

2. Riemann-Roch Theorem on Riemann Surfaces

Let M be a closed Riemann surface. (A Riemann surface is a complex manifold of complex dimension 1. Equip M with the orientation determined by the complex structure.)

Let L be a holomorphic line bundle on M and let

$$\bar{\partial}_L : \ \Gamma(L) \to \Gamma(L \otimes \overline{T^*_{\mathbb{C}} M})$$

be the Dolbeault operator on it. (TM is naturally equipped with a structure of a holomorphic vector bundle. Denote it by $T_{\mathbb{C}}M$. $T^*_{\mathbb{C}}M$ is its dual; cf. §4.2 in Chapter 2.) We prove the following theorem.

Theorem 4.18 (Riemann-Roch Theorem).

$$\dim \operatorname{Ker} \bar{\partial}_L - \dim \operatorname{Coker} \bar{\partial}_L = \deg L + 1 - g_M .$$

We explain the right hand side.

(1) On deg. In general, for a complex line bundle F on M, we define $\deg F$ as follows. If a section f of F intersects the zero section transversely, $(df)_x : (TM)_x \to (F)_x$ is an isomorphism at each zero x of f. The degree $\deg F$ is the difference between the number of x where $(df)_x$ respects the orientation and the number of x where it reverses the orientation. It does not depend on the choice of f.

(2) On the genus g_M of M. We define it by $\deg T_{\mathbb{C}}M = 2-2g_M$ for the moment. (It is a fundamental property of oriented closed surfaces that $\deg T_{\mathbb{C}}M$ is an even integer. However, if we define g_M in this way, it is not necessary to show this property. In fact, we can find that $\deg T_{\mathbb{C}}M$ is an even integer in the proof of Theorem 4.18.)

Exercise 1.

(1) Check that $\deg F$ is independent of the choice of f in the following way: for transversal sections f_0 and f_1, we pick a homotopy f_t in a general position and consider the totality $\{f_\bullet\}$ as a section of the complex line bundle $F \times [0,1]$ on $M \times [0,1]$. Its zero set becomes a compact 1-dimensional manifold with boundary. Consider the number of boundary points.

(2) For complex line bundles F, F_0, F_1, show that $\deg(F_0 \otimes F_1) = \deg F_0 + \deg F_1$ and $\deg \bar{F} = \deg F^* = -\deg F$.

The left hand side of the Riemann-Roch theorem can be paraphrased, in terms of $\mathbb{Z}_2$-graded Clifford modules, as follows.

Fix a Riemannian metric preserved by the complex structure and a Hermitian metric on L. Write

$$W^0 = L, \quad W^1 = L \otimes \overline{T^*_{\mathbb{C}}M}$$

and define the Clifford multiplication on W

$$c_W(v) = \begin{pmatrix} 0 & -v^{\lrcorner} \\ v^{\wedge} & 0 \end{pmatrix}$$

for an element v of $TM \,(\cong T^*M)$. (Note that $\overline{T^*_{\mathbb{C}}M}$ and $T_{\mathbb{C}}M \,(= TM)$ are identified using the Hermitian metric on $T_{\mathbb{C}}M$.) Then $D = \bar{\partial}_L + \bar{\partial}_L^*$ is an operator of Dirac type (cf. §4.2 in Chapter 2).

By Theorem 5.44, which will be proved in Chapter 5 (see also Theorem 5.45 and Remark 5.46 in Chapter 5), we find that $\operatorname{Coker} \bar{\partial}_L \cong \operatorname{Ker} \bar{\partial}_L^*$. Hence the left hand side of the Riemann-Roch theorem is equal to $\operatorname{ind} W$. Since M is a closed manifold, we can choose $h = 0$. In the formulation of the Riemann-Roch theorem, the dimensions of these Ker, Coker are those of complex vector spaces. If we denote by $\operatorname{ind}_{\mathbb{R}} W$ the index given by the dimensions as real vector spaces, we have $\operatorname{ind}_{\mathbb{R}} W = 2 \operatorname{ind} W$.

Then the claim of the Riemann-Roch theorem is paraphrased as follows:

$$\operatorname{ind}_{\mathbb{R}} W = \deg(L^{\otimes 2} \otimes T_{\mathbb{C}} M).$$

We would like to show this claim by choosing an appropriate h and applying the Excision Theorem to the pair $(W, [h])$.

Fix a generic anti-linear homomorphism ϕ from L to $L \otimes \overline{T_{\mathbb{C}}^* M}$. In other words, ϕ is a $\mathbb{C}$-linear homomorphism $\phi : \overline{L} \to L \otimes \overline{T_{\mathbb{C}}^* M}$. Since $T_{\mathbb{C}} M$ and L are equipped with the Hermitian metrics, ϕ is regarded as a section of $L^{\otimes 2} \otimes T_{\mathbb{C}} M$, which is complex linear. By saying that ϕ is generic we mean that it intersects 0 transversally.

From now on, we forget the complex structure on W and regard W as just a real vector bundle. W is equipped with a natural Euclidean metric.

Now, we set

$$h = \begin{pmatrix} 0 & \phi^* \\ \phi & 0 \end{pmatrix}. \tag{4.1}$$

Here, ϕ^* is the transpose of ϕ with respect to the Euclidean metric. We find that the symmetric endomorphism h anti-commutes with the Clifford multiplication from Example 2.18 in Chapter 2. Or it follows from the following lemma.

LEMMA 4.19. *Let A, B be elements of $O(2)$ such that* $\det A = 1$, $\det B = -1$. *Then the following two* 4×4*-matrices*

$$\begin{pmatrix} 0 & -A^* \\ A & 0 \end{pmatrix}, \quad \begin{pmatrix} 0 & B^* \\ B & 0 \end{pmatrix}$$

are anti-commutative. Here A^*, B^* are the transposed matrices of A, B.

PROOF. It follows from $A^* = A^{-1}$, $B^* = B$, $BAB^{-1} = B^{-1}AB = A^{-1}$. □

We compute the index of $(W, [h])$, instead of W, using the Excision Theorem.

REMARK 4.20. This is equivalent to counting the number of small eigenvalues of $\bar\partial_t \bar\partial_t^* + \bar\partial_t^* \bar\partial_t$ for a sufficiently large t, where $\bar\partial_t = \bar\partial_L + t\phi$.

Denote the zeros of ϕ by $x_1, x_2, \ldots, x_{d_+}$, resp. $y_1, y_2, \ldots, y_{d_-}$, where x_i, resp. y_j, are zeros such that the derivative $(d\phi)_{x_i}$ respects, resp. reverses, the orientation. Choose neighborhoods U_i, resp. V_j, of x_i, resp. y_j, small enough so that no other zeros are contained. Then, by the definition of deg, we find that

$$\deg(L^{\otimes 2} \otimes T_{\mathbb{C}} M) = d_+ - d_-$$

and by the Excision Theorem,

$$\mathrm{ind}_{\mathbb{R}}(W, [h]) = \sum_{i=1}^{d_+} \mathrm{ind}_{\mathbb{R}}(W|U_i, [h|U_i]) + \sum_{j=1}^{d_-} \mathrm{ind}_{\mathbb{R}}(W|V_j, [h|V_j]).$$

Hence the Riemann-Roch theorem is reduced to the following lemma.

LEMMA 4.21. $\mathrm{ind}_{\mathbb{R}}(W|U_i, [h|U_i]) = 1, \quad \mathrm{ind}_{\mathbb{R}}(W|V_j, [h|V_j]) = -1.$

PROOF. Choose a coordinate on U_i with x_i corresponding to the origin and perturb $h|U_i$ up to homotopy. Then $(W|U_i, h|U_i)$ turns out to be isomorphic to the restriction of the pair of the 2-dimensional supersymmetric harmonic oscillator to a neighborhood of the origin. Similarly, we find that $(W|V_j, h|V_j)$ is locally isomorphic to the pair of the 2-dimensional supersymmetric harmonic oscillator with the parity of the grading exchanged. From these observations, the claims are reduced to the fact that the index of the supersymmetric harmonic oscillator is 1. The details are left to readers. □

3. Mod 2 Index of Spin Structures on Riemann Surfaces

It is known that there are 4^g spin structures on a closed Riemann surface of genus g (cf. Lemma 4.27). We will see that they are divided into two classes using the mod 2 index and determine how many of them there are in each class. In this section, we give only an outline and skip the detailed computation. We assume fundamental knowledge of topology of surfaces.

3.1. Definition of mod 2 Index. Let M be a closed Riemann surface of genus g. Spin structures on M are in one-to-one correspondence with holomorphic line bundles L such that $L^{\otimes 2} = T^*_{\mathbb{C}}M$ (cf. Chapter 2, §3.2, Exercise 3). Write $K = T^*_{\mathbb{C}}M$.

For L such that $L^{\otimes 2} = K$, we consider the Dolbeault operator $\bar{\partial}_L : \Gamma(L) \to \Gamma(\overline{T^*_{\mathbb{C}}M} \otimes L)$. As we explained in the previous section, $\operatorname{Ker} \bar{\partial}_L$ is finite dimensional.

Using the fact that $L^{\otimes 2} = K$, there exists a natural bilinear form

$$(\cdot, \cdot): \ \Gamma(L) \times \Gamma(\overline{T^*_{\mathbb{C}}M} \otimes L) \to \Gamma(T^*_{\mathbb{C}}M \otimes \overline{T^*_{\mathbb{C}}M}) = \Omega^{1,1}(X) \overset{\int_M}{\to} \mathbb{C}$$

between $\Gamma(L)$ and $\Gamma(\overline{T^*_{\mathbb{C}}M} \otimes L)$. This is non-degenerate, hence, in this sense, we regard the two infinite dimensional vector spaces $\Gamma(L)$, $\Gamma(\bar{K} \otimes L)$ as the dual spaces to each other in a formal sense.

By the integration by parts, we find that

$$(s, \bar{\partial}_L s') + (s', \bar{\partial}_L s) = 0$$

for elements s, s' in $\Gamma(L)$. In other words, the bilinear form $(s, \bar{\partial}_L s')$ on $\Gamma(L)$ is skew-symmetric.

We consider a finite dimensional model of this situation (cf. Chapter 1, Example 1.6). Let E be a finite dimensional vector space and E^* its dual space. Suppose that $f : E \to E^*$ is a skew-symmetric transformation. Then the dimension of $\operatorname{Ker} f$ and the dimension of E have the same parity. In fact, we have the following:

$$\dim \operatorname{Ker} f \equiv \dim E, \quad \bmod 2.$$

In particular, the parity of $\operatorname{Ker} f$ does not depend on the choice of a skew-symmetric transformation f.

Now, we think of $\Gamma(L)$, $\Gamma(L \otimes \overline{T^*_{\mathbb{C}}M})$, ∂_L as infinite dimensional analogs of E, E^*, f, respectively, then the parity of $\operatorname{Ker} \partial_L$ is expected to be invariant under, for example, the deformation of complex structure. We will show that this is eventually the case.

Equip M with a Riemannian metric, which is preserved by the complex structure on M. In the previous section, we equipped L with any Hermitian metric. In this section, however, we use the Hermitian metric on L, which is naturally induced from the Riemannian metric on M based on the property that the double tensor product of L is equal to $T^*_{\mathbb{C}}M$. With respect to this metric, we consider the operator $D = \bar{\partial}_L + \bar{\partial}_L^*$ of Dirac type and the operator D^2 of Laplace type.

We set E_λ the totality of eigenfunctions of $\bar{\partial}_L^* \bar{\partial}_L$, the restriction of D^2 to $\Gamma(L)$ with the eigenvalue λ.

LEMMA 4.22. *The restriction of the skew-symmetric bilinear form* $(s, \bar{\partial}_L s')$ *to* E_λ *is non-degenerate. In particular, the dimension of* E_λ *is even.*

PROOF. Using the isomorphism $L \otimes \bar{L} \cong \mathbb{C}$, which is induced from the Hermitian metric on L, we define the conjugate linear isomorphism ϕ from L to $L \otimes \overline{T^*_{\mathbb{C}} M} = (L \otimes \bar{L}) \otimes \bar{L}$. We define h by equation (4.1) in the previous section, which is anti-commutative with the Clifford multiplication, as in the general argument there. Also $Dh + hD$ is an operator without terms involving the differentiation. We picked the Hermitian metric on L and ϕ in a special way. In this case, we can see, by the expression in local coordinates, that $Dh + hD$ is identically zero. Since ϕ preserves the Hermitian metric, we have $\phi^* = \phi^{-1}$ and find that h is conjugate linear such that $h^2 = \mathrm{id}$. The composition hD is of degree 0. Denote by $\hat{D}$ the restriction of hD to $\Gamma(L)$ to get $\hat{D} = \phi^{-1}\bar{\partial}_L = -\bar{\partial}^*_L \phi$. Now, the bilinear form we defined before satisfies the following:

$$(\hat{D}s, \bar{\partial}_L s) = \int |\bar{\partial}_L s|^2 \, d\,\mathrm{vol}\,. \tag{4.2}$$

Based on these preparations, we prove the claim.

Since $\hat{D}^2 = -\bar{\partial}^*_L \bar{\partial}_L$, the condition for s to be an element of E_λ is that $\hat{D}^2 s = -\lambda s$. Let s be an element of E_λ. Since λ is a (non-negative) real number and $\hat{D}$ is conjugate linear, $s' = \hat{D}s$ also satisfies $\hat{D}^2 s' = -\lambda s'$ and turns out an element of E_λ. If $\lambda \neq 0$ and s is not 0, s' is not 0, either (note that $\hat{D}s' = -\lambda s$). Write equation (4.2) into $(s', \bar{\partial}_L s) = \|s'\|^2$; then it clearly implies that the anti-symmetric bilinear form is non-degenerate on E_λ. □

COROLLARY 4.23. *When we vary the complex structure on* M *continuously, the parity of the dimension* $\operatorname{Ker} \bar{\partial}$ *does not change.*

PROOF. We arrange the Riemannian metric on M varying continuously with the complex structure. Note that the multiplicity of the non-zero eigenvalue of $\bar{\partial}^*_L \bar{\partial}_L$ is always even and that the eigenvalue changes continously with the complex structure (which will be proved in Chapter 5). Then the claim is clear. □

Hence the parity of the dimension of $\operatorname{Ker} \bar{\partial}$ is a topological invariant of the spin structure. This is one of what are called **mod 2 indexes**.

Remember that the notion of the spin structure is defined in terms of the structure of oriented 2-dimensional manifolds without appealing to the complex structure on M (cf. the explanation in the last part of Chapter 2, §3). But, in order to prescribe the complex line bundle L corresponding to L, we need to fix a complex structure (or a Riemannian metric invariant under it). We, however, abuse the terminology and call L the "spin structure" for simplicity.

DEFINITION 4.24. For a spin structure L on a closed 2-dimensional manifold M, we define the **mod 2 index** of L as follows:

$$\operatorname{ind}_{\bmod 2}(L) = \dim \operatorname{Ker} \bar{\partial}_L, \quad \bmod 2\,.$$

We call the spin structure L with $\operatorname{ind}_{\bmod 2}(L) \equiv 0$, resp. $\operatorname{ind}_{\bmod 2}(L) \equiv 1$, a **spin structure of even type**, resp. a **spin structure of odd type**.

Denote by $E_{\leq\mu}(M, L)$ the finite dimensional vector space spanned by all eigenfunctions of $\bar{\partial}_L^* \bar{\partial}_L$ with eigenvalues less than or equal to μ. Then we have

LEMMA 4.25. *For any* $\mu > 0$, $\operatorname{ind}_{\bmod 2}(L) = \dim E_{\leq\mu}(M, L) \bmod 2$ *holds.*

PROOF. It follows from Lemma 4.22. □

REMARK 4.26. Denote by $S_{\mathbb{R}}$ the real vector bundle of rank 2 consisting of all eigenvectors of h, defined in the proof of Lemma 4.22, with the eigenvalue 1. Then $L \oplus L \otimes \overline{T_{\mathbb{C}}^* M}$ is regarded as the complexification of $S_{\mathbb{R}}$ (a section of $S_{\mathbb{R}}$ is a Majorana spinor.) In sum, for a spin structure on a 2-dimensional Riemannian manifold, $W^0 = L$ and $W^1 = L \otimes \overline{T_{\mathbb{C}}^* M}$ are complex conjugate to each other, and their direct sum carries the real structure h. Both the involution, which defines the grading, and the Clifford multiplication are anti-commutative[2] with the real structure h.

3.2. Number of Spin Structures of Even, Resp. Odd, Type. Using the localization argument, we determine the number of spin structures of even type, resp. odd type, on the Riemann surface of genus g. First of all, the total number of spin structures is given by the following

[2]In the case of the Minkowski metric, these properties hold in 4-dimension (differs by 2-dimension from the Riemannian case).

LEMMA 4.27. *There are exactly 4^g spin structures on the Riemann surface of genus g.*

PROOF. The proof is based on some facts on topology of surfaces. Using the facts that $\deg T_{\mathbb{C}}M = 2-2g$ is even, that the isomorphism classes of complex line bundles on M (as smooth complex line bundles) are determined by deg, and that there exists a complex line bundle with any degree, we find that there exists at least one spin structure. For two spin structures L_0, L_1, we have $L_0^{\otimes 2} = L_1^{\otimes 2}$. Thus $L_0 \otimes L_1^*$ is a complex line bundle with the structure group $\{\pm 1\}$. By this correspondence, the set of spin structures is in one-to-one correspondence with the isomorphism classes of complex line bundles with the structure group $\{\pm 1\}$ (once we fix a spin structure as a base point of the set of spin structures). The latter is classified by $H^1(M, \{\pm 1\})$, which consists of 2^{2g} elements. □

REMARK 4.28. The isomorphism classes of complex line bundles with the structure group $\{\pm 1\}$ are in one-to-one correspondence with the isomorphism classes of real line bundles. In general, the "difference" of two spin structures on an oriented manifold is described by a real line bundle, while the difference of two $Spin^c$-structures is described by a complex line bundle, which is classified by $H^2(X, \mathbb{Z})$.

We use the **connected sum** in the following argument. Here, the connected sum is the construction, as follows. Remove a small ball from each of two manifolds of the same dimension and glue them along the boundaries. When the two manifolds are oriented, we glue them in a compatible way with the orientations to get a naturally oriented manifold. The connected sum of the oriented closed surface of genus g_0 and the oriented closed surface of genus g_1 is the oriented closed surface of genus g_0+g_1. From the fact that the diffeomorphism classes of oriented closed surfaces are completely classified by the genus, we obtain the result on the mod 2 index of an oriented closed surface from the mod 2 index in the case of genus 1 and the behavior of mod 2 index under the connected sum.

3.2.1. *The case of genus* 1. We obtain the answer in the case of the torus, i.e., of genus 1, as follows. For a complex number τ with the positive imaginary part, we consider $M = \mathbb{C}/(\mathbb{Z} + \mathbb{Z}\tau)$. $T_{\mathbb{C}}M$ is isomorphic to the trivial line bundle as a holomorphic line bundle. Hence, a spin structure L is given by a holomorphic line bundle such that $L^{\otimes 2}$ is the trivial line bundle.

LEMMA 4.29. *There are exactly* 4 *spin structures on the torus* M, *which are parametrized by* $\mathbb{Z}_2 \times \mathbb{Z}_2$. *Denote by* $L(m_1, m_\tau)$ *the spin structure corresponding to* (m_1, m_τ) *to get the natural identification*

$$\begin{aligned}&\Gamma(L(m_1, m_\tau))\\&= \{s : \mathbb{C} \to \mathbb{C} \mid s(z+1) = (-1)^{m_1} s(z),\ s(z+\tau) = (-1)^{m_\tau} s(z)\}.\end{aligned}$$

In this expression, we have $\bar{\partial}_{L(m_1,m_\tau)} s = \dfrac{\partial}{\partial \bar{z}}$.

Therefore, holomorphic sections of $L(m_1, m_\tau)$ are nothing but holomorphic functions on $\mathbb{C}$ with a certain (twisted) periodic condition. By this periodicity, the holomorphic functions are bounded on $\mathbb{C}$. Hence they must be constants. However, unless $m_1 = m_\tau = 0$, the only constant satisfying the (twisted) periodicity is 0. From this fact, we obtain the following lemma.

LEMMA 4.30.

$$\dim \operatorname{Ker} \bar{\partial}_{L(m_1,m_\tau)} = \begin{cases} 1, & (m_1, m_\tau) = (0,0) \\ 0, & (m_1, m_\tau) = (0,1), (1,0), (1,1). \end{cases}$$

This lemma directly describes the mod 2 index for each spin structure.

3.2.2. *Spin structures on open Riemann surfaces.* In order to study the connected sum, we first need to consider the building block, i.e., the open surface obtained by removing a disk from the closed surface.

Let M be a closed Riemann surface and z the local complex coordinates on a neighborhood U of a point p on M. Here, we assume that $z(p) = 0$ and that U corresponds to $|z| < 1$. We also assume that the function z can be extended to a neighborhood of the closure of U, slightly larger than U.

We fix a Riemannian metric on M, which is preserved by the complex structure and locally isometric to the complex plane via z on U. Define a multi-valued function w on $U \setminus \{p\}$ by $\exp(-w) = -z$. We fix a Riemannian metric on $M \setminus \{p\}$ (independent of the Riemannian metric on M), which is preserved by the complex structure and locally isometric to the complex plane via w on $U \setminus \{p\}$. This is a Riemannian metric with a cylindrical end.

LEMMA 4.31. *The restriction induces a one-to-one correspondence between spin structures on* $M \setminus \{p\}$ *and spin structures on* M.

PROOF. It follows from the isomorphism $H^1(M, \{\pm 1\}) \cong H^1(M \setminus \{p\}, \{\pm 1\})$. Refer to the proof of Lemma 4.27. $\square$

LEMMA 4.32. *Let L be a spin structure on M.*

(1) *Holomorphic sections of L decay exponentially on with respect to the cylindrical metric defined on $M \setminus \{p\}$.*
(2) *If a holomorphic section of L on $M \setminus \{p\}$ is bounded with respect to the cylindrical metric on $M \setminus \{p\}$, it extends to a holomorphic section of L on M.*

PROOF. As $\exp(-w) = -z$, we have the change of variables

$$u(w)\sqrt{dw} = s(z)\sqrt{dz} \iff u = e^{-\frac{w}{2}} s \iff u\frac{1}{\sqrt{-z}} = s. \tag{4.3}$$

Thus, if s is bounded, u decays exponentially as the real part of w tends to be larger and larger. It implies the former statement.

If we assume that u is bounded, $zs(z)$ extends continuously through $z = 0$ by setting the value 0 at $z = 0$. Hence, by Riemann's removal of singularity, it extends holomorphically through $z = 0$, which becomes the zero. It implies that $s(z)$ can extend holomorphically through $z = 0$. The latter statement follows from this fact. $\square$

In particular, if a holomorphic section of L on $M \setminus \{p\}$ is bounded, it automatically decays exponentially. It seems strange at the first glance, but it can be easily seen by setting $w = t + \sqrt{-1}\,\theta$ to express $\bar{\partial}_L^* \bar{\partial}_L$ in the local coordinate on the cylinder $\{\mathrm{Re}\, w \geq 0\}/2\pi\sqrt{-1}\,\mathbb{Z}$. Here, the formal adjoint $\bar{\partial}_L^*$ is defined using the metric on $M \setminus \{p\}$. From the change of variables (4.3), we find that the section of L on the cylinder is written as $u(w)\sqrt{dw}$ with $u(w)$ satisfying the twisted periodicity $u(w + 2\pi\sqrt{-1}\,) = -u(w)$. The essential point, which makes the "strange" thing hold, is that the constant section u must be automatically 0 by the twisted periodicity. To observe the exponential decay, we use the local expression $\bar{\partial}_L^* \bar{\partial}_L(u\sqrt{dw}) = (-\partial_t^2 u - \partial_\theta^2 u)\sqrt{dw}$. Using the Fourier expansion in the direction of θ, the twisted periodicity yields

$$u(t + \theta\sqrt{-1}\,) = \sum_{r \in \mathbb{Z} + \frac{1}{2}} u_r(t) e^{r\theta\sqrt{-1}}$$

(r runs over half integers). Here,

$$(-\partial_t^2 - \partial_\theta^2)u = \sum_{r \in \mathbb{Z} + \frac{1}{2}} \left(-\partial_t^2 u_r(t) + r^2 u_r(t)\right) e^{r\theta\sqrt{-1}}.$$

From this expression, if the eigenfunction $u\sqrt{dw}$ of $\bar{\partial}_L^* \bar{\partial}_L$ on the cylinder is bounded when $t \to \infty$ and the corresponding eigenvalue λ is less than $1/4$, we find that

$$u(t + \theta\sqrt{-1}\,) = \sum_{r \in \mathbb{Z} + \frac{1}{2}} u_r(0) e^{-(\sqrt{r^2 - \lambda})t} e^{r\theta\sqrt{-1}}.$$

Its derivatives of any degree also decay exponentially as $t \to \infty$. In particular, if $\lambda = 0$, it is holomorphic.

Using this exponential decay, we can define the mod 2 index based on the cylindrical Riemannian metric on $M \setminus \{p\}$ and obtain the following lemma.

LEMMA 4.33. *Fix any μ such that $0 < \mu < 1/4$. $E_{\le\mu}(M \setminus \{p\}, L)$ is finite dimensional and its parity of the dimension is equal to the* mod 2 *index of* ind $_{\text{mod } 2}(L)$ *on M.*

PROOF. The finite dimensionality follows from the general theory in Chapter 5 (Corollary 5.28; Remark 5.6). Note that a similar argument to that in Lemma 4.25 works. (If the eigenvalue is less than $1/4$, the eigenfunction decays exponentially. Hence, the contribution from the infinity appearing in the integration by parts is 0, and we can treat this case as the case of closed manifolds.) Thus the parity of $E_{\le\mu}(M \setminus \{p\}, L)$ coincides with the parity of the dimension of the space of holomorphic sections of L on $M \setminus \{p\}$, which are bounded (hence, decays exponentially). The latter is identified with the space of holomorphic sections on M by Lemma 4.32. Its parity is, by the definition, ind $_{\text{mod } 2}(L)$. □

The exponential decay in the above argument is a prototype of the general theory in Chapter 5, §2, Lemma 5.9.

3.2.3. *Spin structures on the connected sum.* Next we study the case of the closed Riemann surface M, which is obtained as the connected sum of two closed Riemann surfaces M_0, M_1.

Let p_0, p_1 be a point on M_0, M_1, respectively. Denote by z_0, z_1 coordinates on neighborhoods U_0, U_1 of p_0, p_1, respectively. Suppose that w_0, w_1 are multi-valued functions such that $\exp(-w_0) = -z_0$, $\exp(-w_1) = -z_1$. In a similar way to the previous subsection, we fix the Riemannian metrics on M_0, M_1, which are flat on some neighborhoods of p_0, p_1 and the Riemannian metrics on $M_0 \setminus \{p_0\}$, $M_1 \setminus \{p_1\}$ with cylindrical ends.

With a parameter $T > 0$, we define the connected sum by gluing

$$M(T) := (M_0 \setminus \{|z_0| < e^{-T}\}) \cup (M_1 \setminus \{|z_1| < e^{-T}\})$$

by the relation $z_0 z_1 = \exp(-T)$. While the diffeomorphism class of $M(T)$ does not depend on T, the complex structures and the Riemannian metrics on $M(T)$, by gluing those on two cylinders, are distinct for different T's as Riemann surfaces (or Riemannian manifolds). In a similar way to Lemma 4.31, we have

LEMMA 4.34. *Spin structures on $M(T)$ are in a one-to-one correspondence with the pairs of a spin structure on M_0 and a spin structure on M_1.*

We can regard $M(T)$ as the Riemannian manifold with a cylindrical end obtained from $M(\infty) := (M_0 \setminus \{p_0\}) \coprod (M_1 \setminus \{p_1\})$ by modifying the end. Set $L(T)$, resp. $L(\infty)$, the spin structure on $M(T)$, resp. $M(\infty)$, which correspond to spin structures L_0, L_1 on M_0, M_1. Since all eigenvalues of $-\partial_\theta^2$ are greater than or equal to $1/4$, hence positive, the "localization" argument, we will explain in Chapter 5, also holds in such cases (cf. Remark 5.6, Chapter 5). The next proposition plays a similar role to that of Proposition 4.14, which was the key in the proof of the Morse inequality (cf. Chapter 5, Theorem 5.40). We assume and use it here.

PROPOSITION 4.35. *Suppose that $0 < \mu < 1/4$ is not an eigenvalue of $\bar{\partial}^*_{L(\infty)} \bar{\partial}_{L(\infty)}$ (with eigenfunctions with the exponential decay). (Such μ exists.) For a sufficiently large T, there holds*

$$\dim E_{\leq\mu}(M(\infty), L(\infty)) = \dim E_{\leq\mu}(M(T), L(T)).$$

The left hand side is written as the sum of the contributions from $M_0 \setminus \{p_0\}$ and from $M_1 \setminus \{p_1\}$.

3.2.4. *The sum formula for the mod 2 index.* Now we find the sum formula of the mod 2 index under the connected sum.

PROPOSITION 4.36. *Let L_0, L_1 be the spin structures on M_0, M_1 corresponding to the spin structure L on $M(T)$, respectively. Then the sum formula* $\operatorname{ind}_{\bmod 2} L \equiv \operatorname{ind}_{\bmod 2} L_0 + \operatorname{ind}_{\bmod 2} L_1$ *holds.*

PROOF. It follows from Proposition 4.35 and Lemma 4.25. ☐

Combined with the discussion in the case of genus 1, we can compute the number of spin structures of even type and odd type, respectively. This is the aim of this section.

THEOREM 4.37 (Riemann). *On a closed Riemann surface of genus g, there are*

$(4^g + 2^g)/2$ spin structures of even type and $(4^g - 2^g)/2$ spin structures of odd type.

PROOF. Suppose that there are $a(g)$ spin structures of even type and $b(g)$ spin structures of odd type on a closed Riemann surface of genus g. Then, by Proposition 4.36 and Lemma 4.34, we have

$$a(g_0 + g_1) = a(g_0)a(g_1) + b(g_0)b(g_1),$$
$$b(g_0 + g_1) = a(g_0)b(g_1) + a(g_1)b(g_0).$$

Hence, we find inductively that $a(g)$ is equal to the sum of coefficients of even powers of z in $(a(1) + b(1)z)^g$ and $b(g)$ is equal to the sum of coefficients of odd powers of z. Substituting ± 1 into z and taking their sum and the difference, we obtain

$$a(g) = \frac{(a(1) + b(1))^g + (a(1) - b(1))^g}{2},$$
$$b(g) = \frac{(a(1) + b(1))^g - (a(1) - b(1))^g}{2}.$$

We substitute $a(1) = 3,\ b(1) = 1$, which is obtained by Lemma 4.30, and we get the desired result. □

4. The Case with a Group Action: Lefschetz Formula

The symmetry by the action of a group often makes the computation of the index simpler. For example, if a pair $(W, [h])$ on X admits the symmetry of a torus action, the index $\operatorname{ind}(W, [h])$ is determined only by the local information around the fixed points of the torus action (see Corollary 4.42).

4.1. Definition of the Index Under the Group Action. First of all, we extend the notion of the index to the general case with an action of a compact group.

Let us assume, here, that X is diffeomorphic to the interior of a compact manifold with boundary. Suppose that a compact Lie group G acts on X smoothly. We can choose a Riemannian metric on X which is invariant under the G-action and has a cylindrical end. We also assume that the G-action is lifted to the $\mathbb{Z}_2$-graded Clifford module W on X and preserves the Hermitian metric and the Clifford multiplication. Moreover, we assume that there exists h, which provides the pair $(W, [h])$, and that h is G-equivariant. We choose h such that it is translationally invariant on the cylinder.

Now, we take the operator D of Dirac type, which is translationally invariant on the cylinder, such that D is G-equivariant, and write $D_t = D + th$ for a sufficiently large t. Since everything in the set-up is compatible with the G-action, the G-action is defined on $\operatorname{Ker} D_t^2$

with the grading preserved. Then the index $\mathrm{ind}_G(W, [h])$ is defined as an element of the representation ring $R(G)$ of G by

$$\mathrm{ind}_G(W, [h]) = [\mathrm{Ker}(D_t^2 | B_X(W^0)] - [\mathrm{Ker}(D_t^2 | B_X(W^1)].$$

We can see that it does not depend on auxiliary data and is determined only by the topological information $(W, [h])$ based on the localization in a similar way to the previous argument. (When we showed the topological invariance of the index, we used the fact that the number of eigenvalues less than a value, which is not an eigenvalue, is invariant under small perturbations. Here, we need the fact that the direct sum of eigenspaces corresponding to those eigenvalues does not change so much under small perturbation of the data. This fact will be established in Theorem 5.41 in Chapter 5.)

Taking it into account, we can see that the index $\mathrm{ind}_G(W, [h])$ is defined for any G-manifold X and any G-invariant pair $(W, [h])$, where X is not necessarily diffeomorphic to the interior of a compact manifold with boundary. The argument is quite parallel to the previous one.

It is sometimes more tractable to study the value of the character for each element g in G rather than to define an element in $R(G)$ in an abstract way. We denote the value by $\mathrm{ind}_g(W, [h])$, which is a complex number:

$$\mathrm{ind}_g(W, [h]) = \mathrm{trace}(g | \mathrm{Ker}(D_t^2 | B_X(W^0)) - \mathrm{trace}(g | \mathrm{Ker}(D_t^2 | B_X(W^1)).$$

An element of the representation ring is characterized by the value of the character for each element of G.

For any element g in G, the closure $\overline{\langle g \rangle}$ of the subgroup generated by g is a commutative Lie subgroup of G. (We can see this from the following fact. Any element in a connected compact Lie group is contained in some maximal torus.) Thus, it is enough to determine $\mathrm{ind}_g(W, [h])$ under the assumption that $G = \overline{\langle g \rangle}$.

4.2. The Case Without Fixed Points. Let X^g be the set of points of X, which are fixed by an element g of G and X^G the intersection of X^g for all $g \in G$. If $G = \overline{\langle g \rangle}$, X^G coincides with X^g.

PROPOSITION 4.38. *Let X be a manifold with a G-action and $(W, [h])$ a pair on X. Suppose that the G-action lifts to the G-action on $(W, [h])$. If X^g is empty, we have* $\mathrm{ind}_g(W, [h]) = 0$.

PROOF. Without loss of generality, we assume that $G = \overline{\langle g \rangle}$. In particular, G is a commutative compact Lie group. For later convenience, we denote $\operatorname{supp} h$ by K. K is disjoint from X^g. (This is obvious, since X^g is empty for the time being.) In this situation, there exists a $\mathbb{Z}_2$-graded finite dimensional unitary representation $R = R^0 \oplus R^1$ of G satisfying the following:

(1) $\operatorname{trace}(g|R^0) \neq \operatorname{trace}(g|R^1)$.
(2) There exists a homomorphism $\underline{h} : X \times R^0 \to X \times R^1$ as G-equivariant vector bundles, which is isomorphic over K.

We will show these facts later. Assume them to define a $\mathbb{Z}_2$-graded Hermitian vector bundle $F = F^0 \oplus F^1$ on X by $F = X \times R$. We define the G-action on F by the diagonal action. Consider a $\mathbb{Z}_2 \times \mathbb{Z}_2$-graded Clifford module $\tilde{W} = F \otimes W$ and define two Hermitian endomorphisms on $\tilde{W}$. Firstly, we define $\tilde{h}_0$ by $\tilde{h}_0 = \epsilon_F \otimes h$, and consider the pair $(\tilde{W}, [\tilde{h}_0])$. The structure of tensors immediately implies the following relation:

$$\operatorname{ind}_g(\tilde{W}, [\tilde{h}_0]) = (\operatorname{trace}(g|R^0) - \operatorname{trace}(g|R^1)) \operatorname{ind}_g(W, [h]). \tag{4.4}$$

On the other hand, we define $\tilde{h}_1$ by

$$\tilde{h}_1 = h_F \otimes \mathrm{id}_W + \epsilon_F \otimes h, \quad h_F = \begin{pmatrix} 0 & \underline{h}^* \\ \underline{h} & 0 \end{pmatrix}.$$

Then we find that $\tilde{h}_1^2 = h_F^2 \otimes \mathrm{id}_W + \mathrm{id}_F \otimes h^2$. Hence $\operatorname{supp} \tilde{h}_1$ is the intersection of $\operatorname{supp} h_F$ and $\operatorname{supp} h$. From the construction, $\operatorname{supp} h_F$ is disjoint from $K = \operatorname{supp} h$, hence the intersection is empty. Therefore, by the Excision Theorem, the index $\operatorname{ind}_g(\tilde{W}, [\tilde{h}_1])$ of the pair $(\tilde{W}, [\tilde{h}_1])$ is 0. More concretely, for the corresponding operator $\tilde{D}_t$ of Dirac type,

$$\tilde{D}_t^2 = \tilde{D}^2 + t(\tilde{D}\tilde{h}_1 + \tilde{h}_1\tilde{D}) + t^2\tilde{h}_1^2$$

is a positive operator, when t is sufficiently large. Therefore $\operatorname{Ker} \tilde{D}_t^2 = 0$.

We can find a linear homotopy between $\tilde{h}_0$ and $\tilde{h}_1$ with the support being compact. Hence, we have

$$\operatorname{ind}_g(\tilde{W}, [\tilde{h}_0]) = \operatorname{ind}_g(\tilde{W}, [\tilde{h}_1]).$$

From these facts, we obtain

$$\operatorname{ind}_g(W, [h]) = \frac{\operatorname{ind}_g(\tilde{W}, [\tilde{h}_1])}{\operatorname{trace}(g|R^0) - \operatorname{trace}(g|R^1)} = 0.$$

□

We give a proof of the fact used in the proof above.

LEMMA 4.39. *Suppose that a commutative compact Lie group G acts on X. For an element g in G and a compact subset K, which does not contain any fixed points of g, there exists a unitary representation R^0, R^1 satisfying the following:*

(1) $\operatorname{trace}(g|R^0) \neq \operatorname{trace}(g|R^1)$.
(2) *There exists a homomorphism $\underline{h} : X \times R^0 \to X \times R^1$ as G-equivariant vector bundles, which is isomorphic over K.*

PROOF. We directly construct F and h_F in the proof of Proposition 4.38. For each point x of K, we denote by G_x the set of all elements of G, which fix x. Then the G-orbit Gx of x is diffeomorphic to G/G_x. Denote by U_x a tubular neighborhood of Gx. Since K is compact, there are finitely many points $x_1, x_2, \dots, x_n$ so that K is covered by U_{x_k}. Since x_k is not a fixed point of g, g does not belong to G_{x_k}. Since G is commutative, G_{x_k} is a normal subgroup; hence, G/G_{x_k} is also a compact Lie group. Let R^1_k be a faithful unitary representation space of G/G_{x_k}. We regard it as a unitary representation space of G via a surjective homomorphism $G \to G/G_x$. Denote by R^0_k the representation space of G, which is R_k with the trivial G-action. Since g does not belong to G_{x_k}, we find that $\operatorname{trace}(g|R^0_k) - \operatorname{trace}(g|R^1_k) \neq 0$.

Define a $\mathbb{Z}_2$-graded G unitary representation space R by the tensor product of $R_k = R^0_k \oplus R^1_k$, i.e,, $R = R_1 \otimes R_2 \otimes \cdots \otimes R_n$. Then we have

$$\operatorname{trace}(g|R^0) - \operatorname{trace}(g|R^1) = \prod_{k=1}^{n} (\operatorname{trace}(g|R^0_k) - \operatorname{trace}(g|R^1_k)) \neq 0.$$

Here we introduce some notations.

- We define the map $\underline{h}_k$ from G/G_x to $\operatorname{Herm}(R_k)$ by
$$\underline{h}_k(g \bmod G_x) = \begin{pmatrix} 0 & g^{-1} \\ g & 0 \end{pmatrix}.$$
- Pick a partition of unity adapted to the open covering by $\{U_{x_k}\}$ and $X \setminus K$ and denote it by $\{\rho^2_k\}$ and ρ^2_∞. (Here we take the square so that the notation is compatible with later chapters. It is not important at this moment.)
- Denote by $\pi_k : U_{x_k} \to G/G_x$ the G-equivariant projection map from the tubular neighborhood U_{x_k} of Gx_k to Gx_k.
- Define the $\mathbb{Z}_2$-graded G-equivariant vector bundle F_k on U_{x_k} by $F_k = X \times R_k$ and the Hermitian endomorphism h_k on it

of degree 1 by $h_k = \rho_k \underline{h}_k \pi_k$. (We extend it by 0 outside of U_{x_k}.) h_k is isomorphic on the subset $\rho_k > 0$.

Now, we define the Hermitian endomorphism h_F, of degree 1, on

$$F = X \times R = F_1 \otimes F_2 \otimes \cdots \otimes F_n$$

by

$$\begin{aligned} h_F = \quad & h_1 \otimes \mathrm{id}_{F_2} \otimes \mathrm{id}_{F_3} \otimes \cdots \otimes \mathrm{id}_{F_n} \\ & + \epsilon_{F_1} \otimes h_2 \otimes \mathrm{id}_{F_3} \otimes \cdots \otimes \mathrm{id}_{F_n} \\ & + \cdots + \\ & + \epsilon_{F_1} \otimes \epsilon_{F_2} \otimes \epsilon_{F_3} \otimes \cdots \otimes h_n. \end{aligned}$$

Then its square satisfies

$$\begin{aligned} h_F^2 = \quad & h_1^2 \otimes \mathrm{id}_{F_2} \otimes \mathrm{id}_{F_3} \otimes \cdots \otimes \mathrm{id}_{F_n} \\ & + \mathrm{id}_{F_1} \otimes h_2^2 \otimes \mathrm{id}_{F_3} \otimes \cdots \otimes \mathrm{id}_{F_n} \\ & + \cdots + \\ & + \mathrm{id}_{F_1} \otimes \mathrm{id}_{F_2} \otimes \mathrm{id}_{F_3} \otimes \cdots \otimes h_n^2. \end{aligned}$$

Then it is positive at each point of K, hence h_F is isomorphic everywhere on K. It is also easy to see that h_F is G-equivariant from its construction. □

4.3. Localization to Neighborhoods of Fixed Points. More generally, if the fixed point set is not empty, we have the following "localization".

THEOREM 4.40. *Let $(W, [h])$ be a pair on the manifold X with the G-action. Suppose that the G-action lifts to $(W, [h])$. Then, for any open subset U containing* $\operatorname{supp} h \cap X^g$, *the value* $\operatorname{ind}_g(W, [h])$ *depends only on the data on the restriction of $(W, [h])$ to U (more precisely, the restriction of W and h, not $[h]$, to U).*

Before giving the proof, we state two corollaries, which follow directly from this theorem.

COROLLARY 4.41. *Suppose that the G-action lifts to the pair $(W, [h])$ on the G-manifold X and that X^g is compact. Pick an arbitrary open subset U containing X^g. Then the value* $\operatorname{ind}_g(W, [h])$ *depends only on the data on the restriction of W to U. In particular, it is independent of the choice of h.*

PROOF. Since X^g is compact. we deform h by a homotopy so that it becomes identically zero on a neighborhood of X^g. Using this fact, Corollary 4.41 clearly follows from Theorem 4.40. □

COROLLARY 4.42. *Suppose that the G-action lifts to the pair $(W, [h])$ on the G-manifold X and that G is a torus. Pick an arbitrary open subset U containing X^G. Then the value $\operatorname{ind}_G(W, [h])$ depends only on the data on the restriction of W and h to U. In particular, $\operatorname{ind}(W, [h]) = \operatorname{ind}_e(W, [h])$ depends only on the data on the restriction of W and h to U.*

PROOF. Since G is a torus, there exist g densely such that $G = \overline{\langle g \rangle}$. Since $\operatorname{ind}_g(W, [h])$ for such a g determines $\operatorname{ind}_G(W, [h])$, Corollary 4.41 clearly follows from Theorem 4.40. □

PROOF OF THEOREM 4.40. Fix a relatively compact open subset U' containing $\operatorname{supp} h \cap X^g$. We first mimic the proof of Proposition 4.38. Here, however, we use the compact subset $K = \operatorname{supp} h \setminus U'$. The support of $\tilde{h}_1$ is not necessarily empty, contrary to the previous case. But we can see that it is a compact subset of U'. Thus, the restriction of the pair $(\tilde{W}, [\tilde{h}_1])$ to U also has a compact support.

From the Excision Theorem, we have $\operatorname{ind}_g(\tilde{W}, [\tilde{h}_1]) = \operatorname{ind}_g(\tilde{W}|U, [\tilde{h}|U])$, hence

$$\operatorname{ind}_g(W, [h]) = \frac{\operatorname{ind}_g(\tilde{W}|U, [\tilde{h}_1|U])}{\operatorname{trace}(g|R^0) - \operatorname{trace}(g|R^1)}. \tag{4.5}$$

At the first glance, the right hand side may seem to depend only on the data on U. However, it is not the case in the strict sense, since $\underline{h}$, which was used in the construction of the pair $(\tilde{W}, [\tilde{h}_1])$, is defined based on the information on X globally.

In addition to $R, \underline{h}$, we pick arbitrary $R', \underline{h}'$ satisfying the following conditions.

(1) $\operatorname{trace}(g|R'^0) \neq \operatorname{trace}(g|R'^1)$.
(2) $\underline{h}'$ is a homomorphism as G-equivariant vector bundles on U, i.e., $\underline{h}' : U \times R'^0 \to U \times R'^1$, which is isomorphic on $\operatorname{supp} h \setminus U'$.

(The restriction of $R, \underline{h}$ to U is an example satisfying these conditions.) In this situation, we use $R', \underline{h}'$ instead of $R, \underline{h}$ to define a pair $(\tilde{W}', [\tilde{h}'_1])$ over U by the same formula as in the definition of the pair $(\tilde{W}, [\tilde{h}_1])$. Then, if the equality

$$\frac{\operatorname{ind}_g(\tilde{W}|U, [\tilde{h}_1|U])}{\operatorname{trace}(g|R^0) - \operatorname{trace}(g|R^1)} = \frac{\operatorname{ind}_g(\tilde{W}', [\tilde{h}'_1])}{\operatorname{trace}(g|R'^0) - \operatorname{trace}(g|R'^1)}$$

is proved, the right hand side depends only on the data on U, hence the proof is completed. The argument is quite parallel to the proof of the previous theorem, and we only give a sketch of the proof.

Consider the tensor product $F'' = F \otimes F'$ as a $\mathbb{Z}_2$-graded vector bundle on U and a linear homotopy between two Hermitian endomorphisms of $h_F \otimes \mathrm{id}_{F'}$ and $\epsilon_F \otimes h_{F'}$ of F''. Take the tensor product of W and F''. Then the above linear homotopy induces a homotopy of the pair. Similar to the argument deriving (4.4), we obtain

$$\begin{aligned}&\mathrm{ind}_g(\tilde{W}|U, [\tilde{h}_1|U])(\mathrm{trace}(g|R'^0) - \mathrm{trace}(g|R'^1))\\ &\quad = (\mathrm{trace}(g|R^0) - \mathrm{trace}(g|R^1))\,\mathrm{ind}_g(\tilde{W}', [\tilde{h}'_1]).\end{aligned}$$

The desired equality follows from this equality. □

4.4. The Case with Finitely Many Fixed Points. Particularly when X^g consists of finitely many points, we compute $\mathrm{ind}_g(W,[h])$ concretely.

The supersymmetric harmonic oscillator (W_E, D_E, h_E) plays a fundamental role. Note that it carries an action of the orthogonal group $O(E)$ of E.

$O(E)$ acts on $\mathrm{Ker}(D_{E,t}^2)$ trivially. Thus we find that

$$\mathrm{ind}_O(W_E, [h_E]) = 1$$

for any element O in $O(E)$. This is the basis of our consideration below.

The origin is a fixed point for any element of $O(E)$. $O(E)$ maps the fiber over 0 to itself. The following formula is also another basis of our consideration:

$$\mathrm{trace}(O|(W_E^0)_0) - \mathrm{trace}(O|(W_E^1)_0) = \det(1 - O).$$

This is because W_E is isomorphic (keeping the parity of the degree) to the exterior algebra of $(TE)_0 \cong E$.

In general, when X^g is compact, Corollary 4.41 implies that $\mathrm{ind}_g(W, [h])$ depends only on the restriction of W to a neighborhood of X^g. Therefore it is enough to compute the index of the pair $(W, [h'])$ on a neighborhood of X^g with a suitably chosen h', which is independent of the original h. In practice, it is useful to make a little effort using the supersymmetric harmonic oscillator. When X^g consists of isolated points, this "effort" is easy.

THEOREM 4.43 (Lefschetz formula in the case of finitely many fixed points). *Suppose that the G-action lifts to the pair $(W, [h])$ on* *the G-manifold X. If the fixed points of an element g of G are finitely many, the following formula holds:*

$$\mathrm{ind}_g(W,[h]) = \sum_{x\in X^g} \frac{\mathrm{trace}(g|(W^0)_x) - \mathrm{trace}(g|(W^1)_x)}{\det(1-g|(TX)_x)}.$$

PROOF. By (4.5) in the proof of Theorem 4.40, the problem reduces to the consideration on a neighborhood of each fixed point, namely, the case that X is an orthogonal representation space E of G and the origin 0 is a unique fixed point of g. In this case, $\det(1-g|E)$ is not 0.

G acts on the supersymmetric harmonic oscillator (W_E, D_E, h_E) on E in a natural way. Let us consider $W\otimes W_E$. (This is the "effort".) By Corollary 4.41, we find that

$$\mathrm{ind}_g(W\otimes W_E,[h\otimes \mathrm{id}_{W_E}]) = \mathrm{ind}_g(W\otimes W_E,[\epsilon_W\otimes h_E]). \tag{4.6}$$

Since W is isomorphic to the product bundle $E\times(W)_0$ as a G-equivariant $\mathbb{Z}_2$-graded vector bundle and W_E is isomorphic to the product bundle $E\times(W_E)_0$, we obtain

$$\begin{aligned}&\mathrm{ind}_g(W,[h])(\mathrm{trace}(g|(W_E^0)_0) - \mathrm{trace}(g|(W_E^1)_0))\\ &\quad = (\mathrm{trace}(g|(W^0)_0) - \mathrm{trace}(g|(W^1)_0))\,\mathrm{ind}_g(W_E,[h_E]).\end{aligned}$$

Note that

$$\mathrm{trace}(g|(W_E^0)_0)-\mathrm{trace}(g|(W_E^1)_0) = \det(1-g|E), \quad \mathrm{ind}_g(W_E,[h_E]) = 1$$

to get the desired formula. □

REMARK 4.44.

(1) The principle of the proof above can also be expressed in the following way. Consider the pair on $E\times E$ defined as the product of $(W,[h])$ and $(W_E,[h_E])$. (4.6) implies that the index of the restriction of the pair to $E\times\{0\}$ coincides with the index of its restriction to $\{0\}\times E$. Noting that two obvious embeddings, $E\cong E\times\{0\}\to E\times E$ and $E\cong\{0\}\times E\to E\times E$, are joined by a homotopy of linear embeddings of E to $E\times E$, this claim is clear. In Chapter 9 of the sequel, we prove the Bott periodicity by an analogous argument.

(2) In the general case, when X^g does not necessarily consist of finitely many points, the index of the pair $(W,[h])$ on X can be calculated in terms of the index of a certain pair on X^g by an extended version of the argument above. We explain this in detail in Chapter 10 of the sequel.

We will use such a phenomenon—that the indices of suitable pairs on the manifold and its closed submanifolds coincide—in the core of the argument in the proof of Chapter 6.

SUMMARY.

4.1 If a $\mathbb{Z}_2$-graded Clifford module with compact support is isomorphic, in a neighborhood of its support, to the supersymmetric harmonic oscillator, we can compute its index concretely using the Excision Theorem.

4.2 The Poincaré-Hopf theorem and the Riemann-Roch theorem for Riemann surfaces are such examples; the indices are computed in such a way.

4.3 As for variants of the index, or more precise information than the index, there are also examples, which can be approached by the localization. The Morse inequality and the mod 2 index for spin structures on Riemann surfaces are such examples.

4.4 In the study under group actions, we reduced the problem to simple cases by taking "products" (twice). In later proofs of the index theorem and the Bott periodicity theorem, we use similar ideas.

CHAPTER 5

Localization of Eigenfunctions of the Operator of Laplace Type

In this chapter, we explain analysis, which is necessary for the study of the index.

1. Set-up

DEFINITION 5.1. Suppose that the quadruple (W, W', D, V) satisfies the following conditions.

- W, W' are real vector bundles with the Euclidean inner product on a Riemannian manifold X.
- $D : \Gamma(W) \to \Gamma(W')$ is a first order linear partial differential operator.
- Denote by $\sigma_1(D) : T^*X \to \mathrm{Hom}(W, W')$ the principal symbol of D. Then $\sigma_1(D)(v)$ is injective for any $v \neq 0$.
- V is a symmetric endomorphism of W.

In this situation, we call the second order partial differential operator $D^*D + V$ on W the **operator of Laplace type**.

Here D^* is the formal adjoint of D.

EXAMPLE 5.2.

(1) Write $W = X \times \mathbb{R}$, $W' = T^*X$ for the Riemannian manifold X. Let D be the differential d acting on functions on X. Then $\sigma_1(d) = \mathrm{id} : T^*X \to \mathrm{Hom}(W, W') = T^*X$ satisfies the conditions in Definition 5.1. Hence the Laplace operator or the Laplacian d^*d is an operator of Laplace type in the sense of Definition 5.1.

(2) For a real Clifford module W, set $W' = W$ and let D be an operator of Dirac type on W. Then D^2 is an operator of Laplace type. In fact, the Clifford multiplication by any non-zero element of T^*X is injective.

(3) Let W be a real Clifford module, $(W, [h])$ a pair and D an operator of Dirac type on W. Then, for any real number t, $(D + th)^2$ is an operator of Laplace type. In fact, $V = t(Dh + hD) + t^2h^2$ is a symmetric endomorphism of W.

REMARK 5.3.

(1) The operator of Laplace type, which we defined above, is the same as what is called the Schrödinger operator in a large number of articles. In this book, the operator of this type arises as the square of the Dirac operator. Taking the origin of the equation into account, it sounds slightly strange to state "the square of the Dirac operator is the Schrödinger operator". Thus we call it the operator of Laplace type[1].

(2) In this chapter, we work with real vector bundles. For a complex vector bundles, we can apply the results in this chapter by forgetting the complex structure.

In this chapter, we study operators of Laplace type on open manifolds. Let X be a Riemannian manifold satisfying Condition (R) (Chapter 3, Assumption 3.22) and $f : X \to \mathbb{R}$ be the proper map appearing in Condition (R) (Assumption 3.22). Write $K_0 = f^{-1}((-\infty, 0])$.

Let (W, W', D, V) be the quadruple, which defines the operator of Laplace type on X. Under the assumption that these behave nicely at the end of X, we study the operator $D^*D + V$ of Laplace type. As the assumption of "nice behavior", we assume the following:

Fix a real number $\bar{\lambda}$ and a positive real number ϵ.

ASSUMPTION 5.4 (Assumption $(\bar{\lambda}, \epsilon)$).

- For each point p on X, there is a point q on K_0 such that the restriction of D to $B_1^\circ(p)$ is isomorphic to the restriction of D to $B_1^\circ(q)$.
- V is a symmetric endomorphism of W and

$$V \geq \bar{\lambda} + \epsilon$$

holds at each point on $f^{-1}([0, \infty))$. Namely, $V - (\bar{\lambda} + \epsilon)\,\mathrm{id}_W$ is positive semi-definite at each point on the cylinder.

In this book, in order to avoid complication, we give an explanation in the case that the end of X is cylindrical and the data, except

[1] The square of the original Dirac operator on the Minkowski space is not the Laplacian but the Klein-Gordon operator.

V, are translationally invariant. The argument in the general case is almost the same. The details are left to readers.

We give the set-up again here.

Let X be a Riemannian manifold with a cylindrical end. The end is assumed to be isometric to $Z \times [0, \infty)$, where Z is a closed Riemannian manifold. We identify the end and $Z \times [0, \infty)$ and call this part the **cylinder**. Denote by K_0 the compact subset, which is the complement of $Z \times (0, \infty)$ in X. K_0 is a compact manifold with boundary Z.

$$X = K_0 \bigcup_Z Z \times [0, \infty).$$

Let (W, W', D, V) be a quadruple, which defines the operator of Laplace type. We assume the following:

Fix a real number $\bar{\lambda}$ and a positive real number ϵ.

ASSUMPTION 5.5 (Assumption $(\bar{\lambda}, \epsilon)'$).

- W, W' are translationally invariant on the cylinder. The Euclidean inner products are also translationally invariant on the cylinder.
- $D : \Gamma(W) \to \Gamma(W')$ is translationally invariant on the cylinder.
- V is a symmetric endomorphism of W such that

$$V \geq \bar{\lambda} + \epsilon$$

holds at each point on the cylinder. Namely, $V - (\bar{\lambda} + \epsilon)\,\mathrm{id}_W$ is positive semi-definite at each point on the cylinder.

Note that we do not require that V is translationally invariant on the cylinder.

REMARK 5.6. Consider the case that the end of X is cylindrical and there exists a formally self-adjoint operator Δ_Z on Z such that $D^*D = -(\partial/\partial l)^2 + \Delta_Z$ holds on the cylinder $Z \times [0, \infty)$. Then we can replace the condition that $V \geq \bar{\lambda} + \epsilon$ in Assumption $(\bar{\lambda}, \epsilon)'$ above by

$$\Delta_Z + V \geq \bar{\lambda} + \epsilon.$$

(In particular, even though $V = 0$, we can apply the argument below if the smallest eigenvalue of Δ_Z is positive.) We used this fact when we studied the Dirac operator associated with the spin structure on open Riemann surfaces in Chapter 4, §3.

In this chapter, we study the eigenvalues and eigenfunctions of D^*D+V. (Honestly speaking, we discuss not functions but sections of W. So we should call them "eigensections". Since this is not a common terminology, we call them "eigenfunctions".)

Firstly, we must decide the asymptotic behavior at the end of the sections we consider. Various set-ups are possible. Here we consider sections as follows.

DEFINITION 5.7. Let B_X be the set of all smooth sections s on W with the following properties:

$$\liminf_{l\to\infty}\int_{Z\times\{l\}}(|s|^2+|(Ds)|^2)=0.$$

We state the properties for the formal adjoint D^*, which we will use later.

LEMMA 5.8. *Let s, s' be smooth sections of W, W', respectively. The integration of (Ds, s') on $Z\times[l_0, l_1]$ is equal to the integration of (s, D^*s') up to the contribution from the boundary. The boundary contribution is expressed in terms of the following bilinear map, which depends only on the principal symbol $\sigma(D)$ of D:*

$$\{\,,\,\}_{(z,l)}:\ W_{(z,l)}\times W'_{(z,l)}\to\mathbb{R}$$

Namely, we have

$$\int_{Z\times[l_0,l_1]}(s,D^*s')-\int_{Z\times[l_0,l_1]}(Ds,s')=\left(-\int_{Z\times l_1}+\int_{Z\times l_0}\right)\{s,s'\}.$$

2. Exponential Decay

From now on, we study eigenfunctions of D^*D+V, which belong to B_X, with the eigenvalue less than or equal to $\bar\lambda$. We show that such eigenfunctions decay exponentially on the cylinder as $l\to\infty$.

Firstly, we study the behavior of eigenfunctions on the cylinder $Z\times[0,L]$ of finite length. Let s be an eigenfunction of D^*D+V on $Z\times[0,L]$ with the eigenvalue less than or equal to $\bar\lambda$.

The conclusion is the following. Suppose that we have upper bounds for the norms of s and Ds on neighborhoods of the two boundary components $Z\times\{0\}$ and $Z\times\{L\}$.

Then, according to the estimates for s and Ds, they decay approximately exponentially, or faster than exponentially, as they move to the interior from each of the boundary components. Such a decay estimate holds until the points, where two graphs of exponential functions from both sides intersect. (If they do not intersect, the bigger one has priority.)

More precisely, we have

LEMMA 5.9 (exponential decay). *There exist a positive constant $L_0 > 0$, which depends only on the bilinear form $\{\,,\}$ and ϵ such that the following estimate holds. Let $L \geq 3L_0$. Suppose that a smooth section s of W on $Z \times [0, L]$ satisfies*

$$(D^*D + V)s = \lambda s$$

for some $\lambda \leq \bar{\lambda}$. Then the following equality holds for any $L_0 \leq l \leq L - 2L_0$:

$$\begin{aligned}
&\int_{Z\times[l,l+L_0]} (|Ds|^2 + (s,(V-\lambda)s)) \\
&\quad \leq \max\left\{ \frac{1}{2^{l/L_0}} \int_{Z\times[0,L_0]} (|Ds|^2 + (s,(V-\lambda)s)), \right. \\
&\qquad\qquad \left. \frac{1}{2^{(L-L_0-l)/L_0}} \int_{Z\times[L-L_0,L]} (|Ds|^2 + (s,(V-\lambda)s)) \right\}.
\end{aligned}$$

REMARK 5.10. Using the a-priori estimate we give later, we find that the derivative of s of any order also decays exponentially.

PROOF. Choose a constant $C_0 > 0$ such that $|\{a, b\}| \leq C_0(|a|^2 + |b|^2)$. C_0 depends only on $\{\,,\}$. Set L_0 by

$$L_0 = 4\max\{C_0, C_0/\epsilon\}.$$

To make the argument clearer, we prove the estimate only in the case that L and l satisfy the following conditions.

- $L = NL_0$ for some integer $N \geq 2$.
- $l = nL_0$ for some positive integer n.

It is enough for later purposes to show the estimate in this case. (For general L and l, we only need to make L_0 a little bigger. This is not difficult and we leave it to readers.)

Let $0 \leq l_0, l_1 \leq L - L_0$ be arbitrary real numbers such that $l_0 + L_0 < l_1$. Then we have

$$\begin{aligned}
0 &= \int_{Z\times[l_0,l_1]} (s, (D^*D + V - \lambda)s) \\
&= \int_{Z\times[l_0,l_1]} (|Ds|^2 + (s,(V-\lambda)s)) + \left(-\int_{Z\times l_1} + \int_{Z\times l_0}\right) \{s, Ds\}.
\end{aligned}$$

We give an upper bound for $\{\,,\,\}$ using C_0:

(5.1)

$$\int_{Z\times[l_0,l_1]}(|Ds|^2+(s,(V-\lambda)s))\le C_0\left(\int_{Z\times l_1}+\int_{Z\times l_0}\right)(|Ds|^2+|s|^2).$$

Since $(s,(V-\lambda)s)\ge\epsilon|s|^2$ holds on the cylinder, we find that

$$\le\max\left\{C_0,\frac{C_0}{\epsilon}\right\}\left(\int_{Z\times l_1}+\int_{Z\times l_0}\right)(|Ds|^2+(s,(V-\lambda)s)).$$

Now we replace l_0, l_1 in the above inequality by l_0+u, l_1+u and integrate it over $0\le u\le L_0$ to get the following estimate:

$$\begin{aligned}&\left(\int_{Z\times[l_1,l_1+L_0]}+\int_{Z\times[l_0,l_0+L_0]}\right)(|Ds|^2+(s,(V-\lambda)s))\\&\ge\frac{1}{\max\{C_0,C_0/\epsilon\}}\int_0^{L_0}du\int_{Z\times[l_0+u,l_1+u]}(|Ds|^2+(s,(V-\lambda)s))\\&\ge\frac{L_0}{\max\{C_0,C_0/\epsilon\}}\int_{Z\times[l_0+L_0,l_1]}(|Ds|^2+(s,(V-\lambda)s))\\&=4\int_{Z\times[l_0+L_0,l_1]}(|Ds|^2+(s,(V-\lambda)s)).\end{aligned}$$

(Here we changed the order of integration and used the fact that the integrand is non-negative.)

For an integer $0\le n<N$, we set

$$a_n=\int_{Z\times[nL_0,(n+1)L_0]}(|Ds|^2+(s,(V-\lambda)s))\ge 0.$$

Then, from the above estimate, we find that

$$\sum_{n_0<n<n_1}a_n\le\frac{1}{4}(a_{n_0}+a_{n_1})\tag{5.2}$$

for any integer $0\le n_0<n_1\le N$.

Using a_n, we can put the estimate to be proved into

$$a_n\le\max\left\{\frac{a_0}{2^n},\frac{a_{N-1}}{2^{N-1-n}}\right\}.$$

Here n is an arbitrary integer such that $0<n<N-1$. We prove this using the inequality (5.2). The idea is to estimate a_n appropriately in an alternative way, from smaller n and larger n. The actual proof is on the induction on the size of N in the following way.

If $N=2$, there is nothing to prove. We study the case that $N>2$. We consider the case that $a_0\le a_{N-1}$. (The case that

$a_0 \geq a_{N-1}$ is handled in a similar way.) Then, from the inequality (5.2), we particularly find that

$$a_{N-2} \leq \frac{1}{4}(a_0 + a_{N-1}) \leq \frac{1}{2} a_{N-1}.$$

On the other hand, if we consider $\{a_n\}$ in the range of $0 \leq n \leq a_{N-2}$, we can apply the assumption of the induction to get

$$a_n \leq \max\left\{\frac{a_0}{2^n}, \frac{a_{N-2}}{2^{N-2-n}}\right\}.$$

Combining these, we obtain the claim in the whole range $0 \leq n \leq a_{N-1}$. □

Next, we consider sections, which are eigenfunctions of $D^*D + V$ on the whole X. From the above lemma and the inequality (5.1) in its proof, we immediately obtain the following

COROLLARY 5.11. *If an element s of B_X satisfies*

$$(D^*D + V)s = \lambda s$$

for some $\lambda \leq \bar{\lambda}$, we have

$$\int_{Z\times[l,l+L_0]} (|Ds|^2 + (s, (V-\lambda)s)) \leq \frac{1}{2^{l/L_0}} \int_{Z\times[0,L_0]} (|Ds|^2 + (s, (V-\lambda)s))$$

for any $l \geq L_0$. In particular, s and Ds are bounded in the L^2-sense on X.

If there exists an eigenfunction s with the eigenvalue less than or equal to $\bar{\lambda}$, it localizes on the union of K_0 and $Z \times [0, L_0]$ in the above sense. Denote this union by K_{L_0}. s decays, at worst, exponentially outside of the compact set K_{L_0}.

3. Preliminaries for the Calculus of Variation

We have not yet discussed whether eigenfunctions exist or not. In the study of the index, we will face the situation where we have to show the existence of eigenfunctions without solving the partial differential equations directly. One possible approach is the calculus of variations. The calculus of variations is a method to show the existence of eigenfunctions as critical points of a variational problem with constraints, which is formulated on a suitable function space with a suitable choice of a norm.

3.1. Choice of Function Spaces. We already introduced B_X as the class of sections of W. The conditions for a section belonging to this class consist of the local condition and the condition about the limit at the infinity of the end. Sections in this class have no global constraints. For instance, the L^2 norm is not necessarily finite. The global constraints are set only after requiring the assumption that the section satisfies the differential equation (e.g., eigenfunctions). Such a class of functions is suitable, when we consider geometric situations possibly by the cut-and-paste construction of manifolds.

However, B_X is not appropriate for formulating the differential equation as a variational problem. We would like to consider a variational problem with constraints with introducing a certain norm on the function space. For such a purpose, we first need to consider a class of sections, the norm of which is finite.

To develop the calculus of variations, we would like to define a convenient norm. For this purpose, we define $\|\cdot\|_{D,V+C,X}$ as follows.

Firstly, let C be a real number such that $V + C - 1$ is positive semi-definite everywhere on X. The existence of such C follows from the fact that $V - (\bar{\lambda} + \epsilon)$ is positive semi-definite on the cylinder.

DEFINITION 5.12. Define the norms for smooth sections s, s' of W, W', respectively, on a measurable subset A in the following way:

$$\|s\|_A^2 = \int_A |s|^2 \in [0,\infty], \tag{5.3}$$

$$\|s'\|_A^2 = \int_A |s'|^2 \in [0,\infty], \tag{5.4}$$

$$\|s\|_{V+C,A}^2 = \int_A (s,(V+C)s) \in [0,\infty], \tag{5.5}$$

$$\|s\|_{D,A}^2 = \|Ds\|_A^2 + \|s\|_A^2 \in [0,\infty], \tag{5.6}$$

$$\|s\|_{D,V+C,A}^2 = \|Ds\|_A^2 + \|s\|_{V+C,A}^2 \in [0,\infty]. \tag{5.7}$$

By the definition, we have

$$\|s\|_A \le \|s\|_{D,A} \le \|s\|_{D,V+C,A}.$$

If A is compact, there exists a constant $C_A > 0$ depending only on the restriction of V to A such that

$$\|s\|_{V+C,A}^2 \le C_A \|s\|_A^2. \tag{5.8}$$

If s is an eigenfunction of D^*D+V with the eigenvalue λ, s is also an eigenfunction of D^*D+V+C with the eigenvalue $\lambda + C$. Hence, in the study of eigenfuctions and eigenvalues, we can use $V + C$ in place of V without loss of generality. In other words, we may assume the following assumptions without loss of generality.

ASSUMPTION 5.13 (Assumption $C = 0$). $V - 1$ is positive semi-definite everywhere on X.

In this chapter we assume that $C = 0$. However, theorems in this chapter hold without this assumption. This assumption makes the expression of the norms, which is used in the formulation of the variational problem, a little simpler and the formulae simpler. This is the only reason why we work under this technical assumption.

Define the function space $B_X^{D,V}$ as follows.

DEFINITION 5.14. $B_X^{D,V}$ is the set of all measurable sections s of W, which satisfies the following conditions with a certain measurable section s' of W' and a sequence of smooth sections $\{s_k\}$ of W.

$$\|s_k\|_{D,V,X} < \infty, \quad \|s'\|_X^2 + \|s\|_{V,X}^2 < \infty, \tag{5.9}$$

$$\lim_{k\to\infty} (\|Ds_k - s'\|_X^2 + \|s_k - s\|_{V,X}^2) = 0. \tag{5.10}$$

Eigenfunctions belonging to B_X with the eigenvalue less than or equal to $\bar{\lambda}$ also belong to $B_X^{D,V}$ by Lemma 5.9.

Note the following

LEMMA 5.15. *s' in the above definition satisfies*

$$\int_X (s', \beta) = \int_X (s, D^*\beta)$$

for any compactly supported smooth section β of W'.

PROOF. The above equality is clear, if we take s_k, Ds_k in place of s, s'. Then the lemma follows from the fact that s_k, Ds_k converge to s, s' in the L^2 sense, respectively. □

By this lemma, if there are two choices of s', the L^2-inner product of their difference and any compactly supported smooth section β is 0. Hence, we see the uniqueness of s'.

Thus we write $s' = Ds$ and set the right hand side in the second equation of (5.9) by $\|s\|_{D,V,X}^2$. In a similar way, we define $\|s\|_{D,V,A}^2$, $\|s\|_{D,A}^2$, etc.

By the completeness of L^2-spaces, we immediately see the completeness of $B_X^{D,V}$ with respect to $\|\cdot\|_{D,V,X}$. Namely,

LEMMA 5.16. *$(B_X^{D,V}, \|\cdot\|_{D,V,X})$ is a Hilbert space.*

(In fact, $B_X^{D,V}$ is eventually defined by the completion.)

Note that elements of $B_X^{D,V}$ may not be twice differentiable, $D^*D + V$ does not necessarily operate on them.

We will show that elements of $B_X^{D,V}$, which enjoy properties close to eigenfunctions, are localized, although they are not really eigenfunctions of $D^*D + V$.

The following lemma on "localization" is quite simple but essential in our argument below.

LEMMA 5.17. *There exists a constant $C_1 > 0$, which depends only on the restriction of V to K_0 and ϵ with the following property. If an element s of $B_X^{D,V}$ satisfies*

$$\|s\|^2_{V,X} \le \bar{\lambda}\|s\|^2_X,$$

$$\|s\|^2_X \le C_1\|s\|^2_{K_0}$$

holds.

PROOF. In the given inequality on X, we study the contribution from K_0 and $Z \times [0, \infty)$ separately. Then we have

$$\|s\|^2_{V,K_0} + \|s\|^2_{V,Z\times[0,\infty)} \le \bar{\lambda}(\|s\|^2_{K_0} + \|s\|^2_{Z\times[0,\infty)}).$$

Since $V \ge \bar{\lambda} + \epsilon$ on $Z \times [0, \infty)$, we find that

$$\|s\|^2_{V,K_0} + \epsilon\|s\|^2_{Z\times[0,\infty)} \le \bar{\lambda}\|s\|^2_{K_0}.$$

Dropping the first term in the left hand side, which is non-negative, we obtain the desired estimate for $C_1 = 1 + \bar{\lambda}/\epsilon$. □

3.2. Local Properties of the Operator of Laplace Type. The following lemma is well known in the local theory of differential operators. Let D be a first order linear differential operator enjoying the properties in Definition 5.1.

LEMMA 5.18. *Let U be an open subset of X and K a compact subset of U.*

(1) (*Gårding's inequality*). *Let W'' be an arbitrary Euclidean vector bundle on U, $\nabla : \Gamma(U, W) \to \Gamma(U, W'')$ any linear partial differential operator with smooth coefficients of order at most 1. Then there exists a constant $C_{U,K,D,\nabla} > 0$*

depending only on U, K, ∇ and the restriction of D to U, such that

$$\|\nabla s\|_K^2 \leq C_{U,K,D,\nabla} \|s\|_{D,U}^2$$

holds for any smooth section s on U of W.

(2) (*Rellich's Theorem*). *Suppose that a sequence of smooth sections $\{s_k\}$ on U satisfies the following boundedness condition. (It is equivalent to what is called the L_1^2-boundedness.)*

- *Let W'' be an arbitrary Euclidean vector bundle on U, $\nabla : \Gamma(U, W) \to \Gamma(U, W'')$ any linear partial differential operator with smooth coefficients of order at most 1. Then, for any compact subset K' of U, $\{\| \nabla s_k \|_{K'}^2\}$ is bounded.*

Then $\{s_k\}$ contains a subsequence, which is L^2-convergent on K.

Combining these two results, if $\{s_k\}$ is a sequence of smooth sections on U with $\{\|s_k\|_{D,U}\}$ being bounded, we can pick a subsequence, which is L^2-convergent on K.

Since $B_X^{D,V}$ is defined as a certain completion of the space of smooth sections, similar properties hold for a sequence of elements in $B_X^{D,V}$ (check them). We will use this fact in the following form.

COROLLARY 5.19. *Let $\{s_k\}$ be a sequence of elements in $B_X^{D,V}$ such that $\{\|s_k\|_{D,V,X}\}$ are bounded. Then there exists a L^2-convergent subsequence on any compact subset K.*

The following estimate is called the a-priori estimate for the operator of Laplace type.

PROPOSITION 5.20 (a-priori estimate). *Let W'' be any Hermitian vector bundle on U, and $\mathcal{D} : \Gamma(U, W) \to \Gamma(U, W'')$ any linear partial differential operator with smooth coefficients of order at most $2k$. Then there exists a constant $C_{U,K,D,\mathcal{D}} > 0$ depending only on $U, K, \mathcal{D}$ and the restriction of D to U such that*

$$\|\mathcal{D}s\|_K^2 \leq C_{U,K,D,\mathcal{D}}(\|(D^*D)^k s\|_U^2 + \|s\|_U^2)$$

for any smooth section on U of W.

These estimates play important roles in the argument below.

REMARK 5.21.

(1) The a-priori estimate and Rellich's theorem are local properties of the operator of Laplace type. It is enough to show them on the Euclidean space. In this book, we use local properties without giving proofs.

(2) The assumption in Rellich's theorem presented above is too strong. It is impossible to check the assumption for all first order partial differential operators, and it is not necessary, either. As for an open subset in the k-dimensional Euclidean space, it is sufficient to check the boundedness for partial derivatives in k directions.

(3) The a-priori estimate does not necessarily hold for arbitrary linear partial differential operators. This reflects a special property of the operator of Laplace type. The property is that the principal symbol is invertible at any non-zero element of T^*X. This property is phrased that the operator of Laplace type is elliptic.

(4) Gårding's inequality for the operator of Dirac type can be proved by Weitzenböck's formula. Then the a-priori estimate is proved by the induction on the order of the differential operator.

(5) We can also establish the a-priori estimate using only the ellipticity. Firstly, we construct a certain operator P using the inverse map of the principal symbol of D^*D in a concrete way. P is constructed as what is called a "pseudo-differential operator of order -2", which is an extension of integral operators. The principal symbol of P is the inverse map of the principal symbol of D^*D. The construction is not difficult. (First of all, we construct it locally and glue them by a partition of unity. The local construction is the composition of (1) the Fourier transform, (2) the multiplication by the principal symbol and (3) the inverse Fourier transform. The order is determined according to the order of growth of the principal symbol. If we work on the Euclidean space, the gluing construction is not necessary.) P is approximately the left inverse of D^*D. Hence we obtain the estimate for s from the estimate for D^*Ds.

(6) Compared with differential operators, the merit of integral operators is that the order of differentiability is increased by applying the integral operator, if its integral kernel enjoys

a nice property. The "pseudo-differential operator of order -2" increases the order of differentiability by 2.

(7) Roughly, an "approximate left inverse" means that the difference of the composition $P \circ (D^*D)$ and the identity becomes a "pseudo-differential operator of order -1". Such a (left) inverse is called a **(left) parametrix**.

(8) On the torus, Rellich's theorem is easily proved using Fourier expansion. The general case is reduced to the case of the torus.

Assuming Lemma 5.18, we would like to prove the proposition, which claims the existence of eigenfunctions.

Before doing this, we see that Gårding's inequality immediately implies that the function space $B_X^{D,V}$ does not depend so much on the choice of the operator $D^*D + V$.

PROPOSITION 5.22. *The space $B_X^{D,V}$ defined based on the operator $D^*D + V$ and the space $B_X^{D,V}$ defined based on $D'^*D' + V'$ are the same, if $V/c < V' < cV$ holds for some $c > 1$: In fact, the following holds for some $c' > 1$:*

$$\frac{1}{c'}\|s\|_{D,V,X} \leq \|s\|_{D',V',X} \leq c'\|s\|_{D,V,X}\,.$$

PROOF. By Gårding's inequality, we find that the norms $\|\cdot\|_{D,X}$ and $\|\cdot\|_{D',X}$ are equivalent. □

In particular, $B_X^{D,V}$ does not depend on the choice of D (but depends on the growth of V on the cylinder).

The strategy for the existence of eigenfunctions in §4 is that we first show the existence of weak solutions and then show that they are solutions in the classical sense. Here is the definition of weak solutions.

DEFINITION 5.23. Let U be an open set in X and V' a smooth symmetric endomorphism of $W|U$, and u an L^2-section of W on U. An L^2-section s on U is called a **weak solution** of the equation $(D^*D + V')s = u$, if s satisfies

$$\int_U ((D^*D\beta, s) + (\beta, V's)) = \int_U (\beta, u)$$

for any smooth section β, the support of which is a compact subset of U.

PROPOSITION 5.24 (Regularity for weak solutions of the equation of Laplace type). *Let U be an open set in X and V' a smooth symmetric endomorphism of $W|U$, and u a smooth section of W on U. Suppose that an L^2-section s on U is a weak solution for the equation $(D^*D+V')s = u$. Then s is a smooth section and a classical solution for the equation.*

REMARK 5.25.

(1) The above regularity is a local property of the operator of Laplace type as well as the a-priori estimate and Rellich's theorem. Thus we use them without giving a proof.
(2) One of the approaches to the proof of the regularity of weak solutions is to use a systematic approximation (Friedrichs' mollifier) of s. Roughly speaking, we can show that s is smooth by applying the a-priori estimate for the approximating sequence and Sobolev's embedding theorem.

4. Calculus of Variations

The starting point of the study of the calculus of variations on a manifold with a cylindrical end is the following property, which is proved by "localization".

PROPOSITION 5.26. *Suppose that a sequence $\{s_k\}$ of elements of $B_X^{D,V}$ satisfies*

$$\|s_k\|_X^2 = 1, \quad \|s_k\|_{D,V,X}^2 \leq \bar{\lambda}.$$

Then, by taking a suitable subsequence, it converges to an element s_∞ in $B_X^{D,V}$ in the weak sense. We also find that $s_\infty \neq 0$ and the following inequality holds:

$$\frac{\|s_\infty\|_{D,V,X}^2}{\|s_\infty\|_X^2} \leq \liminf_{k\to\infty} \|s_k\|_{D,V,X}^2 . \tag{5.11}$$

PROOF. Recall that we can pick a weakly convergent sequence from any bounded sequence in a Hilbert space. Since $\|s_k\|_{D,V,X}$ are bounded, we may assume that the sequence s_k converges to an element s_∞ of $B_X^{D,V}$. It is sufficient to show that the claim holds after taking a suitable subsequence again.

By Corollary 5.19, take a subsequence, if necessary, to assume that $\{s_k\}$ is L^2-convergent on any compact set of X. Its limit coincides with the restriction of s_∞. By Lemma 5.17, we find that

$\|s_k\|_{K_0}^2 \geq C_1^{-1}$. Note also that s_k converges, in the L^2-sense, to s_∞ on K_0. Thus $\|s_\infty\|_{K_0}^2 \geq C_1^{-1} > 0$, hence s_∞ is not 0.

Let $\delta > 0$ be an arbitrary positive real number.

Pick a sufficiently large compact set K containing K_0 to obtain

$$\frac{\|s_\infty\|_{D,V,K}^2}{\|s_\infty\|_K^2} \geq \frac{\|s_\infty\|_{D,V,X}^2}{\|s_\infty\|_X^2} - \frac{\delta}{2}.$$

In the left hand side, we examine the change of the value of the denominator and the numerator separately when we replace s_∞ by s_k with a sufficiently large k.

The sequence of the denominators of s_k is a strongly convergent sequence, but the sequence of the numerators is only a weakly convergent sequence.

It is easy to deal with the denominators, which are in strong convergence. Namely, we have

$$\|s_\infty\|_K^2 = \lim_{k\to\infty} \|s_k\|_K^2.$$

As for the numerators, which are in weak convergence, we use the following fact. In general, for a weakly convergent sequence in a Hilbert space, the norm of the weak limit may become smaller but does not exceed the lower limit of the norm of the sequence. Namely, we have

$$\|s_\infty\|_{D,V,K}^2 \leq \liminf_{k\to\infty} \|s_k\|_{D,V,K}^2.$$

Combining these considerations on the denominators and the numerators, we find that

$$\frac{\|s_k\|_{D,V,K}^2}{\|s_k\|_K^2} \geq \frac{\|s_\infty\|_{D,V,K}^2}{\|s_\infty\|_K^2} - \frac{\delta}{2}$$

for any sufficiently large k.

On the other hand, since K contains K_0, $V \geq \bar{\lambda} + \varepsilon$ holds outside of K. From this, we have

$$\frac{\|s_k\|_{D,V,X\setminus K}^2}{\|s_k\|_{X\setminus K}^2} \geq \bar{\lambda} + \varepsilon.$$

Combining the estimate on K and the estimate on $X \setminus K$ we obtain the estimate on X as follows:

$$
\begin{aligned}
(5.12)\qquad \|s_k\|^2_{D,V,X} &= \frac{\|s_k\|^2_{D,V,K} + \|s_k\|^2_{D,V,X\setminus K}}{\|s_k\|^2_K + \|s_k\|^2_{X\setminus K}} \\
(5.13)\qquad &\geq \min\left\{ \frac{\|s_k\|^2_{D,V,K}}{\|s_k\|^2_K}, \frac{\|s_k\|^2_{D,V,X\setminus K}}{\|s_k\|^2_{X\setminus K}} \right\} \\
(5.14)\qquad &\geq \min\left\{ \frac{\|s_\infty\|^2_{D,V,X}}{\|s_\infty\|^2_X} - \delta, \bar{\lambda} + \varepsilon \right\}.
\end{aligned}
$$

Since the first term is less than or equal to $\bar{\lambda}$ by the assumption, the following inequality holds for any sufficiently large k:

$$\|s_k\|^2_{D,V,X} \geq \frac{\|s_\infty\|^2_{D,V,X}}{\|s_\infty\|^2_X} - \delta .$$

□

Before proceeding further, we note important properties, which follow directly from Proposition 5.26.

DEFINITION 5.27. Denote by E_λ, resp. $E_{\leq\lambda}$, the vector space spanned by eigenfunctions of $D^*D + V$, which belongs to B_X, such that the corresponding eigenvalues are λ, resp. less than or equal to λ. Namely,

$$E_\lambda = \{s \in B_X \mid (D^*D + V)s = \lambda s\}, \quad E_{\leq\lambda} = \bigoplus_{\mu\leq\lambda} E_\mu .$$

By Lemma 5.9, we find that elements of $E_{\leq\lambda}$ belong to $B_X^{D,V}$.

When eigenvalues are less than or equal to $\bar{\lambda}$, we easily see that eigenfunctions belonging to distinct eigenvalues are L^2-orthogonal to each other as in the case of symmetric transformations acting on finite dimensional vector spaces. (Based on the exponential decay of eigenfunctions, we can freely use the integration by parts on X.) From now on, the orthogonality with respect to the L^2-inner product is one of our tools.

COROLLARY 5.28. *If $\lambda \leq \bar{\lambda}$, $E_{\leq\lambda}$ is finite dimensional.*

PROOF. Suppose that it is infinite dimensional. Denote by $\{s_k\}$ an orthonormal basis with respect to the L^2-inner product. Then it satisfies the assumption of Proposition 5.26. On the other hand, the L^2-weak limit of this sequence is 0. Therefore, for any subsequence, the weak limit, if it exists, with respect to $\|\cdot\|_{D,V,X}$ must be 0. It contradicts the conclusion of Proposition 5.26. $\square$

Denote by p_E the orthogonal projection, with respect to the L^2-inner product, from $B_X^{D,V}$ to the finite dimensional subspace E consisting of L^2-bounded smooth sections in $B_X^{D,V}$. In particular, for $\lambda \leq \bar{\lambda}$, $E_{\leq\lambda}$ is finite dimensional. We denote $p_{E_{\leq\lambda}}$ by $p_{\leq\lambda}$. Write $p_{>\lambda} = \mathrm{id} - p_{\leq\lambda}$. Then the following properties hold:

$$s = p_{\leq\lambda}s + p_{>\lambda}s, \quad p_{\leq\lambda}s \in E_{\leq\lambda}, \quad \int_X (p_{\leq\lambda}s_0, p_{>\lambda}s_1) = 0.$$

We state the properties, which are used later, as the following lemma.

LEMMA 5.29.

(1) *Let E be a finite dimensional subspace of $B_X^{D,V}$. If $\{s_k\}$ converges to s_∞ in the L^2-weak sense, $p_E s_k$ converges to $p_E s_\infty$ in E.*

(2) $\|s\|_{D,V,X}^2 = \|p_{\leq\lambda}s\|_{D,V,X}^2 + \|p_{>\lambda}s\|_{D,V,X}^2$.

PROOF. Since E is finite dimensional, the statement (1) immediately follows from the expression of p_E in terms of the L^2-inner product with finitely many elements in a basis of E. It is sufficient to show the statement (2) in the case that $p_{\leq\lambda}s$ is an element of E_λ. In such a case, we find the statement from the integration by parts. $\square$

The existence of eigenfunctions is derived from the following theorem.

THEOREM 5.30. *Let $\mu \leq \bar{\lambda}$. Suppose that a finite dimensional subspace E of $B_X^{D,V}$ consists of smooth sections, contains $E_{\leq\mu}$, and is stable under the operator $D^*D + V$. Moreover, we assume that there exists an element s of $B_X^{D,V}$ such that*

$$\|s\|_X^2 = 1, \quad \|s\|_{D,V,X}^2 \leq \bar{\lambda}, \quad p_E s = 0.$$

(1) *Among these s, there exists an element which attains the infimum of $\|s\|_{D,V,X}^2$.*

(2) *Denote by λ the infimum. Then $\mu < \lambda \leq \bar{\lambda}$ holds.*

(3) *Let s_0 be one of such an s, which attains the infimum. Then s_0 is smooth and satisfies $(D^*D + V)s_0 = \lambda s_0$.*

PROOF. Let $\{s_k\}$ be a sequence in $B_X^{D,V}$ such that

$$\|s_k\|_X^2 = 1, \quad \lim_{k\to\infty} \|s_k\|_{D,V,X}^2 = \lambda, \quad p_E s_k = 0.$$

By taking a subsequence, we assume that it converges in the weak sense with respect to the L^2-norm $\|\cdot\|_X$ and denote its weak limit by s_∞. Then, from Lemma 5.29 (1), we find that $p_E s_\infty = 0$. We further take a subsequence to assume that it also converges to s_∞ in the weak sense with respect to $\|\cdot\|_{D,V,X}$. Proposition 5.26 implies that s_∞ is not 0. Denote its L^2-normalization by $s_0 = s_\infty/\|s_\infty\|_X$. Then it satisfies

$$\|s_0\|_{D,V,X}^2 \leq \lambda, \quad p_E s_0 = 0.$$

Hence, by the choice of λ, the left inequality becomes an equality and s_0 is an element attaining the infimum.

We show that this s_0 is a weak solution of the equation $(D^*D + V - \lambda)s = 0$. For any compactly supported smooth section β, we consider a family of elements of $B_X^{D,V}$, which depends on a real parameter x: $s_x = (s_0 + x(1-p_E)\beta)/\|s_0 + x(1-p_E)\beta\|_X$. The L^2-norm of s_x is 1 and s_x is L^2-orthogonal to E. Then, since λ is the infimum, $\|s_x\|_{D,V,X}^2$ is a smooth function around $x = 0$ and attains the minimum λ at $x = 0$. Hence the derivative of this function at 0 is 0. Manipulating this relation, we obtain

$$\int_X ((Ds_0, D(1-p_E)\beta) + (s_0, V(1-p_E)\beta)) = \lambda \int_X (s_0, (1-p_E)\beta).$$

Since $p_E\beta$ is an element of E, $Dp_E\beta$ decays exponentially. Hence, $D\beta - Dp_E\beta = D(1-p_E)\beta$ also decays exponentially. Using them, we obtain, by the integration by parts,

$$\int_X (s_0, (D^*D + V - \lambda)(1-p_E)\beta) = 0.$$

On the other hand, $p_E\beta$ is an element of E, hence $(D^*D + V)p_E\beta$ is also an element of E. Thus it is L^2-orthogonal to s_0.

$$\int_X (s_0, (D^*D + V - \lambda)p_E\beta) = 0\,.$$

Adding these two equalities, we obtain the condition that s_0 is a weak solution.

From Proposition 5.24, this weak solution is a classical solution and is an eigenfunction of $D^*D + V$ with the eigenvalue λ. Since s_0 and $E_{\leq\mu}$ are orthogonal, we find that $\mu < \lambda$. □

5. Oscillation of Eigenvalues and Eigenfunctions

In this section, we explain the process by which eigenvalues, as well as eigenfunctions, are characterized inductively from smaller ones based on Theorem 5.30.

As for eigenvalues, the following theorem is useful.

THEOREM 5.31. *Suppose that there exists an r-dimensional subspace E of $B_X^{D,V}$, such that $\|s\|_{D,V,X}^2 \leq \bar{\lambda}$ for any element s in E satisfying $\|s\|_X^2 = 1$. Then we have:*

(1) *There are at least r eigenvalues, counted with multiplicities, of $D^*D + V$: $\dim E_{\leq \bar{\lambda}} \geq r$.*
(2) *The infimum of*

$$\max\{\|s\|_{D,V,X}^2 \mid s \in E,\ \|s\|_X^2 = 1\}$$

for all E satisfying the condition above coincides with the r-th eigenvalues counted with multiplicities from the smallest.

PROOF. Suppose that the existence of k eigenvalues less than or equal to $\bar{\lambda}$ for $0 \leq k < r$. We order the k eigenvalues from the smallest and denote them by $\lambda^1, \lambda^2, \ldots, \lambda^k$ with multiplicities. Let E^k be the subspace spanned by these eigenfunctions, which are L^2-orthogonal to each other. Denote by μ the maximal eigenvalue less than λ^k. Since $\dim E > \dim E^k$, there is an element $s \neq 0$ of E, which is L^2-orthogonal to all elements of E^k. Hence, we can apply Theorem 5.30 to E^k and μ. Then s_0 obtained in Theorem 4.5 is an eigenfunction with eigenvalue $\lambda_{k+1} = \lambda$; hence the inductive step is completed. We leave the rest of the claims to readers. □

Once we find such an E satisfying the assumption of Theorem 5.31, we obtain the estimate of eigenvalues.

Under the assumption of the above theorem, there eventually exists such an E, which attains the infimum of $\max\{\|s\|_{D,V,X}^2 \mid s \in E,\ \|s\|^2 = 1\}$. The direct sum of the eigenspaces corresponding to the first up to the k-th eigenvalues attains the infimum. When the r-th eigenvalue coincides with the $r+1$-th, it is clear that E is not unique. Even though all eigenvalues are simple, the space E, which attains the infimum, is not unique. For example, suppose that the first three eigenvalues are simple. Denote the corresponding eigenfunctions by s_1, s_2, s_3 ordered from the smallest eigenvalue. For any x, the absolute value of which is sufficiently small, each $E = \langle s_1 + xs_3, s_2 \rangle$ attains the infimum considered in the above theorem.

Therefore the consideration in the above theorem is not enough to characterize $E_{\le\lambda}$ as a subspace of $B_X^{D,V}$. How can we characterize not only the eigenvalue, but also the eigenspace?

The aim of this section is to find a criterion for a vector space E, which seems "approximately close" to $E_{\le\lambda}$, to be, eventually, "approximately close" to E.

In order to formulate two vector spaces "approximately close", we recall the topology on the Grassmannian manifold. Since we use the L^2-norm $\|\cdot\|_X$, we introduce the following Hilbert space larger than $B_X^{D,V}$.

DEFINITION 5.32. Denote by B_X^2 the set of all L^2-sections of W. It is a Hilbert space with the norm $\|\cdot\|_X$.

Fix a natural number r and denote by Gr_r the set of all r-dimensional subspaces of B_X^2.

DEFINITION 5.33. For elements E, E' of Gr_r, we define $d(E, E') \ge 0$ as the smallest $d \ge 0$ satisfying the following condition: For any orthonormal basis $e_1, e_2, \dots, e_r$ of E, there exists an orthonormal basis $e_1', e_2', \dots, e_r'$ of E' such that $\sum_{k=1}^r \|e_k - e_k'\|_X^2 \le d^2$. An analogous condition holds by interchanging E and E'.

This d is a complete distance on Gr_r. (Check it.) We introduce the topology on Gr_r by this metric.

There are various ways to define the topology on Gr_r. Using the metric d is just one of them. The important point is that this topology makes the subset

$$E_{\mathrm{Gr}_r} = \{(E, s) \mid s \in E\}$$

of $\mathrm{Gr}_r \times B_X^2$ into a (topological) vector bundle on Gr_r in a natural way. Denote by π_{Gr_r} the projection from E_{Gr_r} to Gr_r. We leave the proof of the following lemma as an exercise.

LEMMA 5.34. *For each $E \in \mathrm{Gr}_r$, denote by $\mathcal{U}_E$ the set of all elements E' of Gr_r such that the orthogonal projection from E' to E is an isomorphism as vector spaces. Then it is an open set of Gr_r and $\pi_{\mathrm{Gr}_r} : E_{\mathrm{Gr}_r} \to \mathrm{Gr}_r$ carries a structure of (continuous) vector bundle with the local trivialization*

$$\begin{array}{ccc} \pi_{\mathrm{Gr}_r}^{-1}\mathcal{U}_E & \to & \mathcal{U}_E \times E, \\ (E', s) & \mapsto & p_E s. \end{array}$$

Suppose that a real number λ, λ' satisfies $0 < \lambda < \lambda' \leq \bar{\lambda}$ and the interval $(\lambda, \lambda']$ does not contain eigenvalues of $D^*D + V$. (If we do not put the assumption $c = 0$, we need to replace the first condition by $-C < \lambda < \lambda' \leq \bar{\lambda}$.) Namely, we assume that $E_{\leq\lambda} = E_{\leq\lambda'}$, the dimension of which is denoted by r. Order, with multiplicities, the eigenvalues of D^*D+V less than or equal to λ from the smallest and denote them by $\lambda_1, \lambda_2, \ldots, \lambda_r \leq \lambda$.

In this situation, we give a method to determine whether an r-dimensional subspace E of $B_X^{D,V}$ is located close to $E_{\leq\lambda}$ in Gr_r only in terms of the restriction of the $\|\cdot\|_{D,V,X}$-norm (and the $\|\cdot\|_X$-norm) to E. In this method, the gap $(\lambda, \lambda']$ plays an essential role, as we will see later.

E is regarded as an r-dimensional Hilbert space by the restriction of $\|\cdot\|_X$ to E. In addition, E is also equipped with the symmetric form, which is the restriction of $\|\cdot\|^2_{D,V,X}$ to E. Denote the eigenvalues of the latter by $\lambda_1(E), \lambda_2(E), \ldots, \lambda_r(E)$ in the ascending order. Namely, they are eigenvalues of the matrix, which represents the symmetric form in the orthonormal basis with respect to the former norm.

In a parallel way to Theorem 5.31, we can give a characterization of $\lambda_k(E)$ as follows : $\lambda_k(E)$ is the infimum (minimum) of

$$\max\{\|s\|^2_{D,V,X} \mid s \in E'',\ \|s\|^2_X = 1\}$$

when E'' runs over all k-dimensional subspaces of E.

THEOREM 5.35. *For any $d > 0$, there exists a positive number $\delta_{d,r,\lambda,\lambda',r} > 0$, which depends only on d, r, λ, λ' (continuously) such that the following holds: If an r-dimensional subspace E of $B_X^{D,V}$ satisfies $|\lambda_k(E) - \lambda_k| < \delta_{d,r,\lambda,\lambda'}$ for all $1 \leq k \leq r$, then $d(E, E_{\leq\lambda}) < d$.*

REMARK 5.36. If we do not put the assumption $C = 0$, $\delta_{d,r,\lambda,\lambda',r}$ depends on C also. Precisely, it depends on d, r and the ratio $(C + \lambda')/(C + \lambda)$.

PROOF. Consider the space $\tilde{E}$ spanned by $E_{\leq\lambda}$ and E, which is at most $2r$-dimensional. Then the above proposition is essentially a problem in the finite dimensional $\tilde{E}$. Thus we present only a sketch of the argument. We adopt the induction on r.

Denote by $E_{>\lambda'}$ the orthogonal complement of $E_{\leq\lambda}$ in $\tilde{E}$ with respect to the L^2-norm $\|\cdot\|_X$. By the exponential decay and the integration by parts, we find that it is also orthogonal to $\|\cdot\|^2_{D,V,X}$ in a similar way as in the proof of Theorem 5.30.

Without loss of generality, we assume $\delta_{d,r,\lambda,\lambda'} < \lambda' - \lambda$ to obtain $\lambda_r(E) < \lambda'$. Then the orthogonal projection $p_{\leq\lambda}|E : E \to E_{\leq\lambda}$ is injective. Since the dimensions of both sides are the same, this orthogonal projection is an isomorphism. Denote by $\phi : E_{\leq\lambda} \to E_{>\lambda'}$ the composition of the inverse $(p_{\leq\lambda}|E)^{-1}$ and the orthogonal projection $p_{>\lambda}|E : E \to E_{>\lambda'}$. Then E is identified with the graph of ϕ in $\tilde{E} = E_{\leq\lambda} \oplus E_{>\lambda'}$.

For simplicity, we denote by $\|\cdot\|$ the L^2-norm $\|\cdot\|_X$ on $\tilde{E} = E_{\leq\lambda} \oplus E_{>\lambda'}$ and by $\|\cdot\|^2_{D,V,X}$

$$\|\cdot\|^2_{D,V,X}|\tilde{E} = \|\cdot\|^2_{\leq\lambda} \oplus \|\cdot\|^2_{>\lambda'}$$

the restriction of the square of the norm $\|\cdot\|^2_{D,V,X}$. All eigenvalues of $\|\cdot\|^2_{\leq\lambda}$ are less than or equal to λ and all eigenvalues of $\|\cdot\|^2_{>\lambda'}$ are larger than λ'.

Let e_r be an eigenfunction of the symmetric operator $D^*D + V$ restricted to $E_{\leq\lambda}$ with the eigenvalue λ_r, which is the maximal eigenvalue on $E_{\leq\lambda}$. Suppose that $\|e_r\| = 1$. If an element of $\tilde{E}$ is orthogonal to e_r with respect to $\|\cdot\|^2$, it is also orthogonal with respect to $\|\cdot\|^2_{\leq\lambda} \oplus \|\cdot\|^2_{>\lambda'}$. Taking the norm of the element $e_r \oplus \phi(e_r)$ of E, we find the following inequality:

$$\begin{aligned}\lambda_r(E) &\geq \frac{\|e_r\|^2_{\leq\lambda} + \|\phi(e_r)\|^2_{>\lambda'}}{\|e_r\|^2 + \|\phi(e_r)\|^2} \\ &= \frac{\lambda_r + \|\phi(e_r)\|^2_{>\lambda'}}{1 + \|\phi(e_r)\|^2} \geq \frac{\lambda_r + \lambda'\|\phi(e_r)\|^2}{1 + \|\phi(e_r)\|^2}.\end{aligned}$$

Note the inequality $\lambda_r \leq \lambda < \lambda'$. The last part is the weighted mean of the smaller λ_r and the larger λ'. The weights are determined by the ratio 1 and $\|\phi(e_r)\|^2$. As the latter gets larger, the weighted mean gets larger away from λ_r. If $\lambda_r(E)$ in the first term is "almost" equal to the smaller λ_r, it implies that $\|\phi(e_r)\|^2$ must be "almost" 0. Moreover, in the second part from the last, $\|\phi(e_r)\|^2_{>\lambda'}$ must be also "almost" 0. Here, we mean by "almost" 0 that we may assume that it can be arbitrary small when we choose $\delta_{d,r,\lambda,\lambda',r}$ sufficiently small.

Define $\psi : E_{\leq\lambda} \to E_{>\lambda'}$ by requiring $\psi(e_r) = 0$ and that it is equal to ϕ on the orthocomplement of e_r. Denote the graph of ψ by E'. Then we find, from the fact that $\|\phi(e_r)\|^2$ is "almost" 0, that $d(E, E')$ is also "almost" 0. In particular, $d(E, E_{\leq\lambda})$ and $d(E', E_{\leq\lambda})$ are "almost" equal. Hence it is sufficient to show that $d(E', E_{\leq\lambda})$ is "almost" 0. Taking the fact that $\|\phi(e_r)\|^2_{>\lambda'}$ is "almost" 0 into account, we can show that $\lambda_k(E)$ and $\lambda_k(E')$ are "almost" equal for $1 \leq k \leq r$. Therefore λ_k and $\lambda_k(E')$ are also "almost" equal.

e_r is an eigenfunction of D^*D+V and belongs to the intersection of $E_{\leq\lambda}$ and E'. In particular, one of $\lambda_k(E')$ is equal to λ_r (Remark: It is not necessarily the maximal one among $\lambda_k(E')$. We can construct a counterexample in the case that $\lambda_r = \lambda_{r-1}$.). Hence, by taking the orthocomplements $e_r^{\perp} \cap E_{\leq\lambda}$, $e_r^{\perp} \cap E'$ of e_r, the problem is reduced to the case of dimension $r-1$. □

With these preliminaries, we can state the theorem, which is necessary in the proof of topological invariance of the index.

We consider a family $D_\omega^* D_\omega + V_\omega$ of operators on W parametrized by a topological space Ω.

Here we assume that the family is continuous in the following sense.

(1) For each ω, $D_\omega^* D_\omega + V_\omega$ satisfies the assumption $(\bar{\lambda}, \epsilon)'$.
(2) For each ω, the assumption $C = 0$ holds.
(3) Around each point x, express D_ω, V_ω by the local coordinates and the local trivialization of W on a neighborhood U of x. Then all coefficients are continuous functions on $\Omega \times U$. (These coefficients are differentiable in U-directions, but we do not assume that they are continuous in Ω-directions. Since we do not assume that Ω is a manifold, the derivatives in Ω-directions do not make sense.)
(4) For each ω and each $c > 1$, there is a neighborhood $\mathcal{U}$ of ω such that, for each ω' in $\mathcal{U}$,

$$\frac{1}{c} V_\omega \leq V_{\omega'} \leq c V_\omega$$

holds at all points on the cylinder.

In this situation, the function space $B_X^{D_\omega, V_\omega}$ is independent of ω. We denote this function space simply by $B_X^{D,V}$.

REMARK 5.37. Even though the assumption $C = 0$ does not hold, if the following assumption holds instead of (4), then the function space is independent of ω and we can extend the theorem below. The assumption is: for each ω and each $c > 1$, there is a neighborhood $\mathcal{U}$ of ω such that for each ω' in $\mathcal{U}$,

$$\frac{1}{c}(V_\omega + C_\omega) \leq V_{\omega'} + C_{\omega'} \leq c(V_\omega + C_\omega)$$

holds at all points on the cylinder. Here C_ω is the constant C for each ω.

THEOREM 5.38. *Suppose that, for an element ω of Ω and $\mu < \bar{\lambda}$, $D_\omega^* D_\omega + V_\omega$ has no eigenfunctions in $B_X^{D,V}$ (or equivalently B_X) with the eigenvalue μ. Set by r the dimension of the space $E_{\omega,\leq\mu}$ spanned by eigenfunctions with the eigenvalues less than or equal to μ. Then there exists a neighborhood $\mathcal{U}$ of ω with the following properties:*

(1) *For each element ω' of $\mathcal{U}$, $\dim E_{\omega',\leq\mu} = r$.*
(2) *The following map is continuous:*

$$\mathcal{U} \to \mathrm{Gr}_r, \quad \omega' \mapsto E_{\omega',\leq\mu}.$$

In particular, the totality of $E_{\omega',\leq\mu}$ over $\mathcal{U}$ naturally carries a structure of vector bundle.

PROOF. First of all, using the assumption and Gårding's inequality, we easily find that the norm $\|s\|_{D_\bullet,V_\bullet,X}$ changes continuously in the following sense. Namely, for each $c > 1$, there exists a neighborhood $\mathcal{U}$ of ω such that

$$\frac{1}{c}\|s\|^2_{D_\omega,V_\omega,X} \leq \|s\|^2_{D_\omega,V_{\omega'},X} \leq c\|s\|^2_{D_\omega,V_\omega,X}$$

for any ω' in $\mathcal{U}$ and an element s of $B_X^{D,V}$ with $\|s\|_X = 1$. Pick $c > 1$ so that $c\lambda_{\omega,r} < \mu$. Applying Theorem 5.31 to the operator $D_{\omega'}^* D_{\omega'} + V_\omega$ and $E_{\omega,\leq\mu}$, we find that $\dim E_{\omega',\leq\mu} \geq \dim E_{\omega,\leq\mu}$. On the other hand, applying it to the operator $D_\omega^* D_\omega + V_\omega$ and $E_{\omega',\leq\mu}$, we find that $\dim E_{\omega',\leq\mu} \leq \dim E_{\omega,\leq\mu}$. Combining them, we get the first half of the conclusion. Similarly, for $0 \leq k \leq r$, we find that

$$\begin{cases} \frac{1}{c}\lambda_{\omega,k} \leq \lambda_{\omega',k} \leq c\lambda_{\omega,k} \\ \frac{1}{c}\lambda_{\omega,k}(E_{\omega,\leq\mu}) = \frac{1}{c}\lambda_{\omega,k} \leq \lambda_{\omega',k}(E_{\omega,\leq\mu}) \leq c\lambda_{\omega,k}(E_{\omega,\leq\mu}) = c\lambda_{\omega,k}. \end{cases}$$

Combining them, we obtain

$$\frac{1}{c^2}\lambda_{\omega',k} \leq \lambda_{\omega',k}(E_{\omega,\leq\mu}) \leq c^2\lambda_{\omega',k}.$$

Hence the second half of the conclusion follows from Theorem 5.35. ∎

6. Modification of Operators on Ends

In the assumption of Theorem 5.38, we considered families of operators with the symmetric endomorphisms W, which are of the same order of growth on the cylinder. For the estimate of the norm in functional analysis, such a global assumption is necessary. The next problem is whether the eigenvalues and the space of eigenfunctions do not vary so much, even changing the order of growth of V on the cylinder. Under what kinds of conditions can we claim this?

As a conclusion, the eigenvalues and the eigenspaces do not vary so much when we change V on a region sufficiently far away in the cylinder (whenever Assumption $(\bar{\lambda}, \epsilon)'$ is satisfied).

To show this, we have to study the phenomena which occur on the space X, in addition to the abstract argument on Hilbert spaces. It is, however, not difficult, and the point is the exponential decay of eigenfunctions in Corollary 5.11.

The argument below guarantees not only the freedom of changing V on a region sufficiently far away in the cylinder but also the freedom of modifying the shape of the manifold X itself on a region sufficiently far away in the cylinder.

We consider a way of modifying X and the operator D^*D+V on $Z \times [3L, \infty)$ for $L > 0$.

Suppose that the Riemannian manifold X' with end and boundary and the operator $D'^*D'+V'$ on it satisfy the following conditions.

(1) $V' \geq \bar{\lambda} + \epsilon$ holds everywhere on X'.
(2) The end of X' is isometric to $Z' \times [0, \infty)$ for a certain closed manifold Z' (∞ corresponds to the end).
(3) A neighborhood of the boundary of X' is isometric to $Z \times [3L, 3L + \delta_0)$ for some $\delta_0 > 0$ ($3L$ corresponds to the boundary). Hence, gluing $K_0 \cup Z \times [0, 3L + \delta_0)$ and X' over $Z \times [3L, 3L + \delta_0)$ to get a Riemannian manifold X'' with a cylindrical end.
(4) $D^*D + V$ and $D'^*D' + V'$ are glued over $Z \times [3L, 3L + \delta_0)$ to get an operator $D''^*D'' + V''$ on X''.

REMARK 5.39. The above set-up is in the case of a cylindrical end. If an end is isometric to $Z \times \{x \in \mathbb{R}^r \mid |x| \geq R\}$ in place of $Z \times [0, \infty)$, we take a neighborhood of the boundary of X', which is isometric to $Z \times \{x \in \mathbb{R}^r \mid 3L \leq |x| < 3L+\delta\}$ in place of $Z \times [3L, 3L+\delta)$. Moreover, we can generalize the argument to the general case in almost the same way.

We present the following theorem.

THEOREM 5.40. *Let μ be a real number less than or equal to $\bar{\lambda}$. Suppose that there are no eigenfunctions of $D^*D + V$ with the eigenvalue μ in $B_X^{D,V}$ (or B_X). Then there exists a sufficiently large*

L satisfying the following. Modify the end of X and the operator on it in such a way that the above conditions hold on $Z \times [3L, \infty)$ and construct X'' and $D''^* D'' + V''$. Then the number of eigenvalues less than or equal to μ does not change:

$$\dim E_{\leq \mu}(D^* D + V) = \dim E_{\leq \mu}(D''^* D'' + V'').$$

In fact, we will show the more precise statement not only on eigenvalues but also on eigenfunctions. Since B_X^2 and $B_{X''}^2$ are different function spaces, $E_{\leq \mu}(D^* D + V)$ and $E_{\leq \mu}(D''^* D'' + V'')$ are subspaces of distinct function spaces. Hence, we cannot consider their distance directly. Thus, we pick a smooth function β_L, $\gamma_L : [0, 3L] \to [0, 1]$ satisfying

$$\begin{aligned}
&\beta^2 + \gamma^2 = 1, \\
&\beta_L(l) = 1, \quad \gamma_L(l) = 0 \quad (0 \leq l \leq L), \\
&\beta_L(l) = 0, \quad \gamma_L(l) = 1 \quad (2L \leq l \leq 3L), \\
&0 \geq \beta_L'(l) \geq -1, \quad 0 \leq \gamma_L'(l) \leq 1 \quad (0 \leq l \leq 3L) \\
&|\beta_L''(l)|, |\gamma_L''(l)| \leq 1 \quad (0 \leq l \leq 3L).
\end{aligned}$$

We assume that L is sufficiently large. Multiplying β_L to elements of B_X^2 or $B_{X''}^2$, we obtain sections with the support contained in $K_0 \cup Z \times [0, 2L)$.

We can also regard them as elements of $B_{X''}^2$ as well as B_X^2. Lest any confusion arise, we denote these maps by $\beta_L \cdot$. Symbolically, we have

$$\beta_L \cdot : \; B_X^2, B_{X''}^2 \to B_X^2, B_{X''}^2.$$

THEOREM 5.41. *Let μ be a real number less than or equal to $\bar{\lambda}$. Suppose that there are no eigenfunctions of $D^* D + V$ with the eigenvalue μ in $B_X^{D,V}$ (or B_X). Then, for any $\delta > 0$, there exists a sufficiently large L such that $\beta_L \cdot$ is injective on $E_{\leq \mu}(D^* D + V)$, $E_{\leq \mu}(D''^* D'' + V'')$ and satisfies the following:*

$$\begin{aligned}
d(E_{\leq \mu}(D^* D + V), \beta_L \cdot E_{\leq \mu}(D^* D + V)) &< \delta, \\
d(E_{\leq \mu}(D''^* D'' + V''), \beta_L \cdot E_{\leq \mu}(D''^* D'' + V'')) &< \delta, \\
d(E_{\leq \mu}(D^* D + V), \beta_L \cdot E_{\leq \mu}(D''^* D'' + V'')) &< \delta, \\
d(E_{\leq \mu}(D''^* D'' + V''), \beta_L \cdot E_{\leq \mu}(D^* D + V)) &< \delta .
\end{aligned}$$

PROOF. We first show that the operation of multiplying β is injective on $E_{\leq \mu}(D''^* D'' + V'')$. If an element s of $E_{\leq \mu}(D''^* D'' + V'')$ with L^2-norm 1 is multiplied by β_L to get 0, $V'' > \bar{\lambda}$ on the support of s. Then, $\|s\|_{D'',V'',X''} > \bar{\lambda}$ should hold. Since $\bar{\lambda} \geq \mu$, it contradicts the fact that s is an element of $E_{\leq\mu}(D''^*D'' + V'')$. Hence we obtain the desired injectivity.

Suppose that s is an eigenfunction belonging to $E_{\leq\mu}(D''^*D'' + V'')$ with L^2-norm 1. By applying Lemma 5.9 to it on the cylinder $Z \times [0, 3L]$, we obtain the exponential decay estimate. In particular, if L is sufficiently large, $\|s\|_{D'',V'',Z\times[L,2L]}$ is "almost" equal to 0. Note that there is an orthonormal basis of $E_{\leq\mu}(D''^*D'' + V'')$ with respect to the L^2-metric, which consists of eigenfunctions. Hence, even though s is not necessarily an eigenfunction, $\|s\|_{D'',V'',Z\times[L,2L]}$ is "almost" equal to 0 for s in $E_{\leq\mu}(D''^*D'' + V'')$ of L^2-norm 1.

From this, we find that the following equalities "almost" hold for s in $E_{\leq\mu}(D''^*D'' + V'')$ of L^2-norm 1:

$$\|s\|^2_{X''} = \|\beta_L s\|^2_{X''} + \|\gamma_L s\|^2_{X''}, \tag{5.15}$$

$$\|s\|^2_{D'',V'',X''} = \|\beta_L s\|^2_{D'',V'',X''} + \|\gamma_L s\|^2_{D'',V'',X''}. \tag{5.16}$$

Equality (5.15) clearly holds. A similar equality also holds for $\|\cdot\|^2_{V'',X''}$. As for equality (5.16), it follows from the following identity of operators. (Apply both sides to s and take the L^2-inner product with s itself.)

$$\begin{aligned} D''^*D'' &= \beta_L D''^*D''\beta_L + \gamma_L D''^*D''\gamma_L \\ &\quad + \frac{1}{2}\left([\beta_L, [\beta_L, D''^*D'']] + [\gamma_L, [\gamma_L, D''^*D'']]\right). \end{aligned}$$

Take the commutators twice to get the operators $[\beta_L, [\beta_L, D''^*D'']]$, $[\gamma_L, [\gamma_L, D''^*D'']]$, which do not involve differentiation and are 0 outside of $Z \times [L, 2L]$. These two commutators only depend on the first derivatives of β_L, γ_L (the terms involving derivatives of second order are canceled out.) From the assumption on the estimate of the size of the derivatives of β_L, these two commutators are bounded at each point on $Z \times [L, 2L]$. (The error terms may be positive and/or negative.)

We will show that $\|\gamma_L s\|^2_{X''}$ and $\|\gamma_L s\|^2_{D'',V'',X''}$ are "almost" equal to 0 later.

Write r for the dimension of $E_{\leq\mu}(D''^*D'' + V'')$. From (5.15) and (5.16), we can show that

$$\lambda_k(E_{\leq\mu}(D''^*D'' + V'')) = \lambda_k(\beta \cdot E_{\leq\mu}(D''^*D'' + V'')) + (\text{error term})$$

for $1 \leq k \leq r$. Similarly, the following equality also "almost" holds:

$$\lambda_k(E_{\leq\mu}(D^*D + V)) = \lambda_k(\beta \cdot E_{\leq\mu}(D^*D + V)) + (\text{error term}).$$

On the other hand, note that the following inequalities hold by Theorem 5.31:

$$\begin{aligned} \lambda_k(E_{\le\mu}(D''^*D''+V'')) &\le \lambda_k(\beta\cdot E_{\le\mu}(D^*D+V)),\\ \lambda_k(E_{\le\mu}(D^*D+V)) &\le \lambda_k(\beta\cdot E_{\le\mu}(D''^*D''+V'')).\end{aligned}$$

Combining them, we find that these inequalities are, in fact, "almost" equalities. Hence Theorem 5.41 follows from Theorem 5.35. In addition, we obtain Theorem 5.40 from Theorem 5.41.

The remaining task is the estimate of the size of $\gamma_L s$. It is sufficient to show the following two inequalities:

$$\|\gamma_L s\|_{D'',V'',X''} \ge (\bar\lambda+\epsilon)\|\gamma_L s\|^2_{X''}, \tag{5.17}$$

$$\|\gamma_L s\|_{D'',V'',X''} \le \mu\|\gamma_L s\|^2_{X''} + (\text{error term}). \tag{5.18}$$

Inequality (5.17) clearly follows from the fact that $V'' \ge \bar\lambda+\epsilon$ holds on $\operatorname{supp}\gamma_L s$. As for inequality (5.18), we find it using the following. (Apply both sides to s and take the L^2-inner product with $\gamma_L s$.)

$$(D''^*D''+V'')\gamma_L = \gamma_L(D''^*D''+V'') + [D''^*D'',\gamma_L].$$

The second term in the right hand side is a first order differential operator involving the derivatives up to the second order of γ_L on $Z\times[L,2L]$. By the exponential decay estimate for s and Gårding's inequality (Lemma 5.18 (1)), the L^2-norm of the image of s by the second term of the right hand side is "almost" equal to 0. Note that the eigenvalue of s is less than or equal to μ to obtain (5.18). □

REMARK 5.42. Theorem 5.40 can be extended to more general cases satisfying Assumptions (R), (D) than the Riemannian manifolds with cylindrical ends. For the precise formulation, we only need heavy notation and leave it as an exercise. By the way, we also used Theorem 5.40 in the case of a Euclidean-like end in the proof of the Morse inequality. Other theorems in this chapter can be extended in a similar way.

7. The Case of Closed Manifolds: Spectral Decomposition

We have discussed, up to now, the case of manifolds with cylindrical ends. All these arguments can be applied to the case of closed manifolds (it is the case that Z is empty). In this case, we do not need to think of localizations and the exponential decay in calculus of variations, hence the argument is considerably simplied. $\bar\lambda$ can be taken as an arbitrary large real number.

THEOREM 5.43. *Let $D^*D + V$ be an operator of Laplace type on a closed Riemannian manifold X.*

(1) *The set of all eigenvalues are bounded from below and discrete without accumulation points. The multiplicity of each eigenvalue is finite.*
(2) *The space of all L^2-sections carries an orthonormal basis consisting of (smooth) eigenfunctions.*

PROOF. The former statement is Corollary 5.28 in the case of closed manifolds. The latter is shown as follows. Suppose that there is a section which is L^2-orthogonal to all eigenfunctions. Then it contradicts Theorem 5.30. □

From the above theorem, we find the following.

THEOREM 5.44. *Let W be a vector bundle with a Euclidean metric on a closed manifold X. Suppose that a first order differential operator Q on W satisfies $Q^2 = 0$ and $D = Q + Q^*$ is an operator of Laplace type. Then the following linear map is defined and it is injective:*

$$\operatorname{Ker} D \longrightarrow \frac{\operatorname{Ker} Q}{\operatorname{Im} Q}, \qquad s \mapsto [s]$$

PROOF. Since $Q^2 = (Q^*)^2 = 0$, we have $D^*D = D^2 = QQ^* + Q^*Q$, which commutes with Q, Q^*. Hence Q, Q^* preserves each eigenspace of D^2. By the integration by parts, we find that $\operatorname{Ker} D = \operatorname{Ker} D^2 = \operatorname{Ker} Q \cap \operatorname{Ker} Q^*$. In particular, the above linear map is surely defined. (It also implies that the above linear map is injective. Thus the essential part of the theorem is the surjectivity.)

In general, for a smooth section s satisfying $Qs = 0$, we decompose it into eigen components with "distinct" eigenvalues by Theorem 5.43 (2) and write $s = \sum_{k=0}^{\infty} s_k$. This sum is an L^2-orthogonal decomposition and $D^2 s_k = \lambda_k s_k$, $Qs_k = 0$ holds. Moreover, λ_k diverges to infinity as k tends to infinity. Without loss of generality, we assume that $\lambda_0 = 0$. (If 0 is not an eigenvalue, we set $s_0 = 0$ in the argument below.) Then we complete the proof by showing that

$$s' = \sum_{k=1}^{\infty} \frac{Q^* s_k}{\lambda_k}$$

is L^2-convergent, the limit is smooth and

$$s = s_0 + Qs'.$$

The L^2-convergence follows from

$$\|s'\|_X^2 = \sum_{k=1}^{\infty} \frac{\|Ds_k\|_X^2}{\lambda_k^2} = \sum_{k=0}^{\infty} \frac{\|s_k\|_X^2}{\lambda_k} < +\infty.$$

The smoothness follows from Proposition 5.24 by noting that s' is a weak solution of $Ds' = s - s_0$ (or $D^2 s' = D(s - s_0)$). □

As a special case of this theorem, the following **Hodge-Kodaira theorem**, due to Hodge-Kodaira, is important.

THEOREM 5.45 (Hodge-Kodaira).

(1) *Let X be a closed Riemannian manifold and θ a local system with a Euclidean metric on X. For each element of the cohomology $H^\bullet(X, \theta)$ of the de Rham complex $(\Gamma(\theta \otimes \bigwedge^\bullet T^*X), d)$ with coefficients in θ, there exists a unique harmonic form representing it. Namely,*

$$\operatorname{Ker}\left(dd^* + d^*d \,\middle|\, \Gamma\left(\theta \otimes \bigwedge\nolimits^{\bullet} T^*X\right)\right) \cong H^\bullet(X, \theta).$$

(2) *Let X be a closed Kähler manifold and F a holomorphic vector bundle with a Hermitian metric on X. For each element of the cohomology $H^{0,\bullet}(X, F)$ of the Dolbeault complex $(\Gamma(F \otimes \bigwedge^\bullet \overline{T_{\mathbb{C}}^* X}), \bar\partial)$ with coefficients in F, there exists a unique harmonic form (with respect to the Laplacian associated with the Dolbeault operator) representing it. Namely,*

$$\operatorname{Ker}\left(\bar\partial\bar\partial^* + \bar\partial^*\bar\partial \,\middle|\, \Gamma\left(F \otimes \bigwedge\nolimits^{\bullet} \overline{T_{\mathbb{C}}^* X}\right)\right) \cong H^{0,\bullet}(X, F).$$

REMARK 5.46. Let f be a function on X. We can apply Theorem 5.44 to the case that the exterior differentiation d, resp. the Dolbeault operator $\bar\partial$, is replaced by their conjugate $e^{tf} d e^{-tf}$, resp. $e^{tf} \bar\partial e^{-tf}$. Thus the statement of Theorem 5.45 also holds in such cases. For the same reason, when X is a Riemann surface and the length of the Dolbeault complex is 2, the statement also holds when the Dolbeault operator is replaced by any first order differential operator with the same symbol as the Dolbeault operator.

SUMMARY.

5.1 For an operator of Laplace type with "good behavior" on the end, the eigenfunctions with small eigenvalues decay exponentially.

5.2 Under the same assumption, the space spanned by eigenfunctions with small eigenvalues is finite dimensional. It is characterized with the help of the calculus of variations.
5.3 Under the same assumption, the small eigenvalues do not change so much when we modify the operator of Laplace type as well as the end in a sufficiently far away region.
5.4 Under the same assumption, the finite dimensional space spanned by eigenfunctions with small eigenvalues do not chage so much as a space in the above modification.

CHAPTER 6

Formulation and Proof of the Index Theorem

In Chapter 4, we proved Poincaré-Hopf's theorem and Riemann-Roch's theorem. In this chapter, we explain the principle of our method of computing the index for a wider class of differential operators. A practical method will be given in Chapter 8 of the sequel. In Chapter 4, we considered the case that solutions are localized to a part of a manifold. In this chapter, the point is to study the problem after embedding the given manifold to a bigger manifold. A "bigger manifold" is the Euclidean space of sufficiently large dimension.

1. Strategy for Our Formulation and Proof

The goal of this chapter is to give a procedure to determine the index by topological information. The objects, of which we study the index, are the operators of Dirac type associated with compactly supported $\mathbb{Z}_2$-graded Clifford modules $(W, [h])$.

Unless otherwise specified, a pair means a compactly supported $\mathbb{Z}_2$-graded Clifford module as before.

Define the infinite dimensional unitary group U as follows. For $n < m$, we regard $U(n)$ as a closed subgroup of $U(m)$ by the embedding

$$A \mapsto \begin{pmatrix} A & O \\ O & I_{m-n} \end{pmatrix}$$

from $U(n)$ to $U(m)$ and take the union of $U(n)$ with all positive integers n. We denote it by U. We equip U with the topology as the inductive limit. Namely, a subset K of U is a closed subset if and only if $K \cap U(n)$ is a closed subset of $U(n)$ for any n.

In this chapter, we formulate the index theorem in the following way.

THEOREM 6.1 (Index Theorem). *For a given pair* $(W, [h])$ *on a manifold* X, *we can define an element of the homotopy group* $\pi_{2N-1}(U)$ *with a sufficiently large* N *by a topological procedure. The index* $\mathrm{ind}(W, [h])$ *is determined by this element only.*

Here are additional comments.

- The method to construct the element of $\pi_{2N-1}(U)$ is explained in §2 and §5.1.
- By Bott's periodicity theorem, $\pi_{2N-1}(U)$ is isomorphic to $\mathbb{Z}$. As a conclusion, this isomorphism exactly gives the index. The statement of the index theroem as the formula of the index is completed with the description of this isomorphism. We do it, as well as the proof of Bott's periodicity theorem, in Chapter 9 of the sequel.

REMARK 6.2.

(1) It is easy to see that $\pi_{2N-1}(U) = \pi_{2N-1}(U(m))$ if $m \geq N$ (by applying the homotopy exact sequence for fiber bundles). The reason why we introduced the infinite dimensional unitary group U is just to make the description simpler.
(2) If $N < M$, a natural homomorphism $\pi_{2N-1}(U) \to \pi_{2M-1}(U)$ is constructed by the **product** of the **Bott class** (Bott element). Using the inductive limit of this homomorphism, we can "hide" the expression "a sufficiently large N" from the statement of the index theorem (as we hide finite dimensional unitary groups by using U). Bott's periodicity theorem claims that this homomorphism is an isomorphism.

Just before defining the element of $\pi_{2N-1}(U)$, we first define the index of a pair on $\mathbb{R}^{2N}$, which is equal to the index of the original pair $(W, [h])$.

The proof of the index theorem consists of three parts.

(1) Starting from a given pair, construct a pair on a Euclidean space of sufficiently large even dimension.
(2) Prove that the index of the constructed pair is equal to the index of the original pair.
(3) Construct an element of the homotopy group of U from the pair on the Euclidean space of even dimension and prove that the index depends only on this element.

The construction in Step (1) is a purely topological procedure. We explain it in §2.

The essential step is the proof of the coincidence of the indices in (2). It consists of the consideration of the product and the consideration with the Excision Theorem. We do it in §3 and §4, respectively.

Step (3) is done in §5.

A more detailed outline of the proof is as follows.

In §2, we explain the following construction.

- Preliminary: Let $(W_X, [h_X])$ be a pair on a manifold X. Embed X into a Euclidean space E of sufficiently large even dimension.
- Construction 1: Construct the pair $(W_\nu, [h_\nu])$ on the total space of the normal bundle ν of $X \subset E$.
- Construction 2: Construct the pair $(W_E, [h_E])$ on the Euclidean space E.

We give the proof of the following statements in §3 and §4, respectively.

- Proof 1: Proof of $\operatorname{ind}(W_X, [h_X]) = \operatorname{ind}(W_\nu, [h_\nu])$.
- Proof 2: Proof of $\operatorname{ind}(W_\nu, [h_\nu]) = \operatorname{ind}(W_E, [h_E])$.

In §5, we discuss the following on the Euclidean space of even dimension.

- Construction 3: Construct the element $[f]$ of the homotopy group of U of odd degree out of the pair $(W_E, [h_E])$ on the even dimensional Euclidean space E.
- Proof 3: Prove that $\operatorname{ind}(W_E, [h_E])$ depends only on $[f]$.

The proof of Theorem 6.1 will be completed through these steps.

REMARK 6.3.

(1) The case when the method of this chapter is directly applied is only this kind of first order elliptic partial differential operator (of Dirac type). It is mostly sufficient for applications to global analysis, but there are some cases when we need to compute the index of elliptic partial differential operators of order greater than or equal to 2.

(2) The index in this chapter means the one defined in §1 and §2 in Chapter 3 that is the integer obtained from the dimension of the solution spaces. There are some variants of the index, hence variants of the index theorem. Those variants of the index do not necessarily take values in integers. However,

the study in the chapter is also applicable to those variants. We explain it in detail in Chapter 10 of the sequel.

(3) In order to compute the index by the procedure given in this chapter, we need terminology to describe how vector bundles are twisted. We would like to obtain a formula computing the index of the operator of Dirac type on a Clifford module from the twisting data of the Clifford module, namely, the terminology to write down the formula. The terminology is provided by characteristic classes in Chapter 7.

2. Construction of the Pair on the Euclidean Space

2.1. Preliminary Step (Embedding into the Euclidean Space). Let $(W_X, [h_X])$ be a pair on X. (In order to remember the base space X, we put X in the suffix.) Without loss of generality, we assume that X is diffeomorphic to the interior of a compact manifold with boundary. Fix a Riemannian metric with cylindrical end on X. (When X is a closed manifold, we pick an arbitrary Riemannian metric and fix it.)

Let D_X be an operator of Dirac type on W_X. Choose W_X, D_X, h_X so that they are translationally invariant. We would like to study the index of $D_{X,t} = D_X + th_X$ for a sufficiently large $t > 0$.

By Corollary 2.4 in Chapter 2, X can be embedded into the Euclidean space E of sufficiently large even dimension. We fix such an embedding. (Here we do not require that it be an embedding as a Riemannian manifold.) Denote by $\pi : \nu \to X$ the normal bundle of X in E. A tubular neighborhood U of X is diffeomorphic to the total space of ν.

2.2. Construction 1 (Construction of the Pair on the Total Space of ν). When the fiber of $\nu \to X$ is $\mathbb{R}^m$, the total space of ν is a twisted product of X and $\mathbb{R}^m$. Recall that X is equipped with the pair $(W_X, [h_X])$ and that there exists the pair associated with the supersymmetric harmonic oscillator on $\mathbb{R}^m$.

When ν is not twisted, we can construct the product of these pairs as explained in Chapter 3, §4. We would also like to construct the pair $(W_\nu, [h_\nu])$ on the total space of ν as the "twisted product" of these pairs in the case that ν is twisted.

We explain it in five steps.

(1) Construction of a family of supersymmetric harmonic oscillators.

We regard the pairs $(W_{(\nu)_x}, [h_{(\nu)_x}])$ associated with the supersymmetric harmonic oscillator on the fiber $(\nu)_x$ at x of X as a smooth family parametrized by x. Take the union of them to get a pair

$$(W_{\text{fiber}}, [h_{\text{fiber}}])$$

on $(\nu, T_{\text{fiber}}\nu)$. Here $T_{\text{fiber}}\nu$ is the vector bundle consisting of tangent vectors along fibers of ν. In the case that X is not a closed manifold, the support of h_{fiber} is not compact. Thus it is not precisely a pair.

(2) Tensor product of the family of supersymmetric harmonic oscillators by coefficients.

Taking the tensor product of $(W_{\text{fiber}}, [h_{\text{fiber}}])$ with $\pi^* W_X$ as coefficients to get a pair

$$(\pi^* W_X \otimes W_{\text{fiber}}, [\epsilon_{W_X} \otimes h_{\text{fiber}}])$$

on $(\nu, T_{\text{fiber}}\nu)$. In the case that X is not a closed manifold, the support of $\epsilon_W \otimes h_{\text{fiber}}$ is not compact. Thus it is not precisely a pair.

(3) Lifting of the pair $(W_X, [h_X])$.

Pullback a given pair $(W_X, [h_X])$ on (X, TX) by π to get a pair

$$(\pi^* W_X, [\pi^* h_X])$$

on $(\nu, \pi^* TX)$. However, the support of $\pi^* h_X$ is not compact, in general. Thus it is not precisely a pair.

(4) Construction of the $(W_\nu, [h_\nu])$.

$\pi^* W_X$, which is used as coefficients in (2), carries a structure of a $\mathbb{Z}_2$-graded Clifford module on $(\nu, \pi^* TX)$. Hence the tensor product

$$W_\nu = \pi^* W_X \otimes W_{\text{fiber}}$$

carries a structure of a $\mathbb{Z}_2$-graded Clifford module on $(\nu, \pi^* TX \oplus T_{\text{fiber}}\nu)$. Write

$$h_\nu = \pi^* h_X \otimes \text{id}_{W_{\text{fiber}}} + \epsilon_{W_X} \otimes h_{\text{fiber}}.$$

Once we show that the support of h_ν is compact, we obtain the pair $(W_\nu, [h_\nu])$. We prove that $\operatorname{supp} h_\nu = \operatorname{supp} \pi^* h \cap \operatorname{supp} h_{\text{fiber}}$. Then it is compact. Firstly, we observe that

$$h_\nu^2 = \pi^* h_X^2 \otimes \text{id}_{W_{\text{fiber}}} + \text{id}_{W_X} \otimes h_{\text{fiber}}^2.$$

Note that h_X, h_{fiber} is Hermitian. For an element u of W_ν, apply both sides of this equality to u and take the Hermitian inner product of them with u to obtain

$$|h_\nu u|^2 = |\pi^* h_X u|^2 + |h_{\text{fiber}} u|^2.$$

Thus we find that

$$\operatorname{Ker} h_\nu = \pi^* \operatorname{Ker} h_X \otimes \operatorname{Ker} h_{\text{fiber}}$$

at each point of ν. It means that $\operatorname{supp} h_\nu = \operatorname{supp} \pi^* h \cap \operatorname{supp} h_{\text{fiber}}$. (This is the argument, which we already used in Chapter 4, §4, etc.)

(5) Regard $(W_\nu, [h_\nu])$ as a pair on $(\nu, T\nu)$.

Pick a splitting $T\nu = \pi^* TX \oplus T_{\text{fiber}}\nu$ of the natural homomorphisms

$$0 \to T_{\text{fiber}}\nu \to T\nu \to \pi^* TX \to 0.$$

Using it, we can regard $(W_\nu, [h_\nu])$ as a pair on $(\nu, T\nu)$.

2.3. Construction 2 (Construction of the Operator on the Euclidean Space E). Identifying the total space of ν and a tubular neighborhood U of X, we can think that the pair $(W_\nu, [h_\nu])$ is defined on the open subset U of the Euclidean space E. We would like to construct a pair on the whole E from this pair. The idea is the following. For a given pair, take the direct sum with an "approximately trivial pair" (in some sense) to make it "really trivial" on the complement of a compact subset. Then we can extend the direct sum trivially outside of U. Concretely, the construction consists of four steps.

(1) Expression of W_ν using the spinor bundle on E.

Fix an orientation of E. E is an even dimensional manifold and carries a unique spin structure. Denote by S_E the spinor bundle associated with this spin structure, then S_E is $\mathbb{Z}_2$-graded. Then, by Proposition 2.39 (1) in Chapter 2, there exists a $\mathbb{Z}_2$-graded Hermitian vector bundle $F = F^0 \oplus F^1$ on U such that

$$W_\nu = F \otimes (S_E | U).$$

h_ν is expressed, using some Hermitian endomorphisms h_F of F, as

$$h_\nu = h_F \otimes \operatorname{id}_{S_E|U} .$$

(2) Deformation of h_F.

We prove that, after deforming h_F by homotopy, $h_F^2 = \operatorname{id}_F$ holds outside some compact subset K of U. Recall that, for a Hermitian transformation h_F, the condition $h_F^2 = \operatorname{id}_F$ is equivalent to that h_F preserves the Hermitian metric (unitarity).

Let $\rho_0 : U \to [0,1]$ be a smooth function with compact support, which is equal to 1 in the neighborhood of points where h_F is not invertible. Write $\rho = 1 - \rho_0$. Denote by h the $(1,0)$-component of h_F. We find that

$$h_F = \begin{pmatrix} 0 & h^* \\ h & 0 \end{pmatrix}.$$

Then, replace h_F by

$$\begin{pmatrix} 0 & h^*(hh^*)^{-\rho/2} \\ (hh^*)^{-\rho/2}h & 0 \end{pmatrix}.$$

It enjoys the required property. (At points where f is not invertible, we use the original h_F.)

Denote by K the support of ρ_0. Outside of K, we have an identification between F^0 and F^1, which respects the Hermitian metrics, through the (deformed) h_F.

(3) Take the direct sum and deform it to a trivial vector bundle on $U \setminus K$.

By Theorem 2.5 in Chapter 2, there exists a Hermitian vector bundle F'^0 on U such that $F^0 \oplus F'^0$ is isomorphic to the trivial vector bundle $U \times \mathbb{C}^r$. For $F'^1 = F'^0$, $F' = F'^0 \oplus F'^1$, set the Hermitian automorphism $h_{F'}$ of F' of degree 1 by

$$h_{F'} = \begin{pmatrix} 0 & \mathrm{id}_{F'^0} \\ \mathrm{id}_{F'^0} & 0 \end{pmatrix}.$$

Then the $\mathbb{Z}_2$-graded Hermitian vector bundle $F \oplus F'$ defined on U and the Hermitian endomorphism $h_F \oplus h_{F'}$ on it are isomorphic to the $\mathbb{Z}_2$-graded Hermitian vector bundle, which is trivial outside of K, and the trivial Hermitian endomorphism on it:

$$(U \setminus K) \times (\mathbb{C}^r \oplus \mathbb{C}^r), \quad \begin{pmatrix} 0 & \mathrm{id}_{\mathbb{C}^r} \\ \mathrm{id}_{\mathbb{C}^r} & 0 \end{pmatrix}.$$

Thus, we extend them trivially outside of U to define the $\mathbb{Z}_2$-graded Hermitian vector bundle F_E on E and the trivial Hermitian endomorphism h_{F_E} on it.

(4) Construction of the pair $(W_E, [h_E])$.

Using them, we define the pair $(W_E, [h_E])$ on E by

$$W_E = F_E \otimes S_E, \quad h_E = h_{F_E} \otimes \mathrm{id}_{S_E}.$$

3. Invariance of the Index: Proof 1 (The Index of the Product)

In this section, we prove that $(W_\nu, [h_\nu])$ has the same index as $(W_X, [h_X])$, namely, Proof 1 in the outline in §1. The construction of the pair $(W_\nu, [h_\nu])$ on ν in Construction 1 in §2.2 is a purely topological procedure and only its homotopy class is uniquely determined. But, in order to compare the indices, the homotopy class is not sufficient, and we need to construct the operator D_ν of Dirac type on W_ν and the Hermitian endomorphism h_ν on it.

The principal part of Proof 1 is to pursue each procedure from (1) to (5) in §2.2 not just topologically but as the concrete construction of the operators of Dirac type.

The process of taking the product in (2) can be carried out in the level of operators of Dirac type directly.

If $\nu \to X$ is an untwisted trivial vector bundle, the process of lifting in (3) is also reduced to a simple process, taking the tensor product with something, even in the level of operators of Dirac type, as in the process (2). However, ν is a twisted vector bundle, in general. Thus we need some thoughts to carry out (3). This is the technical point in Proof 1.

Before going into details, we give a sketch of the processes from (1) to (5) in §2.2.

(1) Construction of the family of supersymmetric harmonic oscillator.

First of all, each fiber of ν over X is isomorphic to $\mathbb{R}^m$. Consider the fiberwise supersymmetric harmonic oscillator $(W_{\mathbb{R}^m}, D_{\mathbb{R}^m}, h_{\mathbb{R}^m})$. By gathering them somehow, we can construct the operator $(\tilde{W}_{\text{fiber}}, \tilde{D}_{\text{fiber}}, \tilde{h}_{\text{fiber}})$.

Note that the identification of the fiber with $\mathbb{R}^m$ is not unique but with ambiguities in $O(m)$. However, since the supersymmetric harmonic oscillator is $O(m)$-invariant, we find that the resulting operator is independent of the choice of the identification. This is the point of this construction.

(2) Tensor product of the family of supersymmetric harmonic oscillators by coefficients.

Moreover, from this operator, obtained by gathering supersymmetric harmonic oscillators, we construct the operator $(\tilde{D}_{\text{fiber}}, \tilde{h}_{\text{fiber}})$ on $\pi^* W_X \otimes W_{\text{fiber}}$ in a mostly trivial way, like taking the tensor product with the identity map of $\pi^* W_X$.

(3) Lifting of (D_X, h_X).

Next, we would like to construct the operator $\tilde{D}_X$, which may be called the lift of D_X. Since the vector bundle ν is twisted, this construction contains ambiguities. Cover X by finitely many open sets, on which ν is trivial. Pick a trivialization on each open set and fix a partition of unity subordinate to this open covering. Then, using these data, we can construct $\tilde{D}_X$ as one of the operators, which may be called the lift of D_X. The obtained operator depends on the choices of local trivializations of ν and a partition of unity. We also construct $\tilde{h}_X$, the lift of h_X. The construction of $\tilde{h}_X$ immediately follows.

REMARK 6.4. If X is a closed manifold, we can take $h_X = 0$. In this case, $\tilde{h}_X = 0$.

(4) Construction of (D_ν, h_ν) and its properties.

For simplicity, we abbreviate by ϵ_{W_X} the operator $\epsilon_{\pi^* W_X} \otimes \mathrm{id}_{W_{\text{fiber}}}$ on $\pi^* W_X \otimes W_{\text{fiber}}$.

Then the desired pair of operators is defined as follows.

DEFINITION 6.5.

$$\begin{aligned} D_\nu &= \epsilon_{W_X} \circ \tilde{D}_{\text{fiber}} + \tilde{D}_X, \\ h_\nu &= \epsilon_{W_X} \circ \tilde{h}_{\text{fiber}} + \tilde{h}_X. \end{aligned}$$

The first terms in the right hand sides are written as the composition of mappings. The right component of the composition comes from the supersymmetric harmonic oscillator and the left component comes from W_X. Each composition is commutative. From the process of the construction, $\tilde{D}_X$ and $\tilde{D}_{\text{fiber}}$ are also commutative.

In the process (5), we will show that D_ν is the operator of Dirac type, which behaves nicely at the end of ν. Taking it for granted, we describe the index of the pair $(W_\nu, [h_\nu])$ using D_ν, h_ν. For $t > 0$, we find from the construction that the index of the operator $D_{\nu,t} = D_\nu + t h_\nu$ is equal to the index of $D_{X,t} = D_X + t h_X$. In fact, we obtain the isomorphism as $\mathbb{Z}_2$-graded vector spaces

$$\operatorname{Ker} D_{\nu,t}^2 \cong \operatorname{Ker} D_{X,t}^2.$$

(We set up appropriate function spaces as the domains of those operators, respectively.)

(5) Regard D_ν as an operator of Dirac type on ν.

We compute the principal symbol of D_ν. Then we will find that D_ν can be regarded as an operator of Dirac type, when we define an appropriate Riemannian metric on the total space of ν. From the construction, D_ν and h_ν in Definition 6.5 behave nicely at the end of ν. Thus, using them, we find that we can compute the index of the pair $(W_\nu, [h_\nu])$ as we did in (4).

In the following subsections, we explain (1) through (5) in order.

3.1. Construction of the Family of Supersymmetric Harmonic Oscillators. Recall that the supersymmetric harmonic oscillator is a first order differential operator acting on the $\mathbb{Z}_2$-graded Clifford module $W_{\mathbb{R}^m} = \bigwedge^\bullet T\mathbb{R}^m$ on $\mathbb{R}^m$.

For technical reasons, we use, in this chapter, the supersymmetric harmonic oscillator $(W_{\mathbb{R}^m_g}, D_{\mathbb{R}^m_g}, h_{\mathbb{R}^m_g,v})$ with a cylindrical end, which was constructed in Chapter 4, §1.3 instead of the original supersymmetric harmonic oscillator. Since we use only this substitute in this chapter, we denote it simply by $(W_{\mathbb{R}^m}, D_{\mathbb{R}^m}, h_{\mathbb{R}^m})$ and call it the supersymmetric harmonic oscillator. Note that the Riemannian metric g on $\mathbb{R}^m$ is not flat but has a cylindrical end.

The operator, which we eventually study, is the one written as $D_{\mathbb{R}^m,t} = D_{\mathbb{R}^m} + th_{\mathbb{R}^m}$ for a sufficiently large parameter $t > 0$.

Using the principal $O(m)$-bundle P_ν associated with ν, ν and the vector bundle $T_{\text{fiber}}\nu$ on ν are expressed as follows:

$$\begin{aligned} \nu &= P_\nu \times_{O(m)} \mathbb{R}^m, \\ T_{\text{fiber}}\nu &= P_\nu \times_{O(m)} T\mathbb{R}^m \end{aligned}$$

Now we define the $\mathbb{Z}_2$-graded Clifford module W_{fiber} on $(\nu, T_{\text{fiber}}\nu)$ by

$$W_{\text{fiber}} = P_\nu \times_{O(m)} W_{\mathbb{R}^m}.$$

Then define the differential operator $\tilde{D}_{\text{fiber}}$ acting on the vector bundle W_{fiber} on ν and the Hermitian endomorphism $\tilde{h}_{\text{fiber}}$ of the vector bundle as follows.

DEFINITION 6.6. Sections of W_{fiber} are regarded as $O(m)$-equivariant maps from P_ν to $\Gamma(W_{\mathbb{R}^m})$. Composing such a map with $D_{\mathbb{R}^m}$, resp. $h_{\mathbb{R}^m}$, we obtain an $O(m)$-equivariant map from P_ν to $\Gamma(W_{\mathbb{R}^m})$. Thus it can be considered as a section of W_{fiber}. We denote the map defined by the composition as above by $\tilde{D}_{\text{fiber}}$, resp. $\tilde{h}_{\text{fiber}}$.

3.2. Tensor Products of the Family of Supersymmetric Harmonic Oscillators by Coefficients. We would like to define the linear differential operator $\tilde{D}_{\text{fiber}}$ acting on the vector bundle $\pi^* W_X \otimes W_{\text{fiber}}$ on ν and the Hermitian endomorphism $\tilde{h}_{\text{fiber}}$ of the vector bundle.

Let us consider whether the construction enjoying the following property is possible.

Pick a section $\pi^* a$ of $\pi^* W_X$, i.e., the pull-back of a section a of W_X and any section $\tilde{b}$ of W_{fiber}. The desired property is

$$\tilde{D}_{\text{fiber}}(\pi^* a \otimes \tilde{b}) = \pi^* a \otimes \tilde{D}_{\text{fiber}} \tilde{b}, \quad \tilde{h}_{\text{fiber}}(\pi^* a \otimes \tilde{b}) = \pi^* a \otimes \tilde{h}_{\text{fiber}} \tilde{b}.$$

Any element of the domain of these operators is written as a finite sum of sections of the form $\pi^* a \otimes \tilde{b}$. (Locally, it is clear by the use of the local triviality of W_X. Using a partition of unity, we can also show that it holds globally. However, the local observation is sufficient to define the desired operator.) Thus, if there is an operator $\tilde{D}_{\text{fiber}}$, $\tilde{h}_{\text{fiber}}$ with the above property, it is completely characterized by this property. The problem is whether the operator is well defined by this property. Namely, fixing a local trivialization of the vector bundle involved in the construction and using the corresponding coordinates, we specify the way of expressing any element in the domain as a finite sum of sections in the form $\pi^* a \otimes \tilde{b}$. Then we can define $\tilde{D}_{\text{fiber}}$, $\tilde{h}_{\text{fiber}}$, which enjoys the above property. At this point, we need to check whether the above property holds for any other expression than the expression which is used in the definition.

In fact, it is straightforward to check this claim by writing it down. Details are left to readers. The point is $\tilde{D}_{\text{fiber}}$, $\tilde{h}_{\text{fiber}}$ at the starting point enjoys the following property: for any smooth function φ on X,

$$\pi^*(\varphi a) \otimes \tilde{D}_{\mathbb{R}^m} \tilde{b} = \pi^* a \otimes \tilde{D}_{\text{fiber}}(\varphi \tilde{b}), \quad \pi^*(\varphi a) \otimes \tilde{h}_{\mathbb{R}^m} \tilde{b} = \pi^* a \otimes \tilde{h}_{\text{fiber}}(\varphi \tilde{b}).$$

The second equality is clear, since $\tilde{h}_{\text{fiber}}$ does not contain derivatives. Since $\tilde{D}_{\text{fiber}}$ only contains the derivatives in the fiber directions and φ is constant along fibers, the first equality holds.

3.3. Lifting of D_X. We would like to define the differential operator $\tilde{D}_X$ acting on sections of the vector bundle $\pi^* W_X \otimes W_{\text{fiber}}$ on ν and the Hermitian endomorphism $\tilde{h}_X$ of the vector bundle.

In §3.2, we started from an operator acting on sections of the vector bundle on ν and constructed a differential operator acting on sections of $\pi^* W_X \otimes W_{\mathrm{fiber}}$. Now we would like to do a similar construction starting from an operator D_X acting on sections of a vector bundle on X.

Mimic the method in §3.2 and consider whether we get a construction with the following property.

Pick a section $\pi^* a$ of $\pi^* W_X$, i.e., the pull-back of a section a of W_X and any section $\tilde{b}$ of W_{fiber}. The first candidate of the desired property is

$$\tilde{D}_X(\pi^* a \otimes \tilde{b}) = \pi^* D_X a \otimes \tilde{b} \quad (?), \quad \tilde{h}_X(\pi^* a \otimes \tilde{b}) = \pi^* h_X a \otimes \tilde{b} \quad (?).$$

If there exist such operators with this property, we can see that it is uniquely determined, as in the case of §3.2.

As for $\tilde{h}_X$, we find that it is well defined by this property, as in the previous case. However, it is not the case for $\tilde{D}_X$. The essential reason for this is the following. For a smooth function φ on X, $\varphi D_X a$ and $D_X(\varphi a)$ do not coincide, in general, hence

$$\pi^* D_X(\varphi a) \otimes \tilde{b} \neq \pi^* D_X a \otimes \varphi \tilde{b}.$$

But we will eventually encounter the situation where we hope to use the above property for $\tilde{D}_X$ in a later section. How should we overcome this dilemma?

In fact, we need not use the above property for arbitrary a and arbitrary $\tilde{b}$. We will find that it is sufficient for our argument that the above property holds only for quite special $\tilde{b}$.

Thus, we construct a lift of D_X in a bit ad-hoc way and try to show that it enjoys the desired property.

If ν is not a twisted vector bundle but of the product form $X \times \mathbb{R}^m$, we may construct an operator on ν by gathering D_X just as we constructed the operator on ν by gathering the supersymmetric harmonic oscillators (with cylindrical ends). Hence, at least on an open subset of X, where ν is trivial, we have such a construction. By gluing them, some operator may be constructed on ν.

In fact, we will show that such an operator constructed by gluing enjoys a suitable property used later.

The concrete procedure is as follows. To avoid inessential complication, we explain it in the case of a closed manifold X. In the case of a general X with a non-empty cylindrical end, we follow the construction below carefully in such a way that the constructed object is translationally invariant on the cylinder. It is not difficult and we leave it to readers.

There are finitely many suitable open subsets $\{U_\alpha\}$ such that X is covered by them and the restriction of ν to each U_α is trivial.

Fix a trivialization. Then we have

$$\nu|U_\alpha \cong U_\alpha \times \mathbb{R}^m.$$

Denote the projection to the second factor by $\pi_\alpha : \nu|U_\alpha \to \mathbb{R}^m$. Using this product structure, the vector bundle obtained by restricting $\pi^* W_X$ to this part is identified with the product of the vector bundle $W_X|U_\alpha$ on U_α and $\mathbb{R}^m$. Namely, the copies of $W_X|U_\alpha$ parametrized by points of $\mathbb{R}^m$ are gathered to obtain it. Then D_X naturally acts on sections of this restricted vector bundle in the direction of U_α. We denote this action by

$$D_{X,\alpha} : \ \Gamma(U_\alpha \times \mathbb{R}^m, \pi^* W_X) \to \Gamma(U_\alpha \times \mathbb{R}^m, \pi^* W_X).$$

Using the product structure, we can also regard the vector bundle as follows:

$$\pi^* W_{\text{fiber}}|(U_\alpha \times \mathbb{R}^m) = \pi_\alpha^*(W_{\mathbb{R}^m}).$$

Based on the preparation above, we define the lift $\tilde{D}_{X,\alpha}$ of D_X with the following property in the case of the direct product $U_\alpha \times \mathbb{R}^m$. Namely, the required property is that

$$\tilde{D}_{X,\alpha}(\tilde{a} \otimes \pi_\alpha b) = D_{X,\alpha}\tilde{a} \otimes \pi_\alpha b$$

for an arbitrary section $\tilde{a}$ of $\pi^* W_X$ and an arbitrary section b of $W_{\mathbb{R}^m}$. The proof is similar to the argument in §3.2. (This case is rather easier thanks to the product structure.)

Let $\{\rho_\alpha^2\}$ be a partition of unity associated with the open covering $\{U_\alpha\}$. Namely, ρ_α is a smooth non-negative function with support in U_α such that $\sum_\alpha \rho_\alpha^2 = 1$. (Here, we consider the square of ρ_α for a later necessity.) Then the composition $\rho_\alpha \tilde{D}_{X,\alpha} \rho_\alpha$ is regarded as an operator defined everywhere on ν. We put $\tilde{D}_{X,\alpha}$ between two ρ_α's to obtain a formally self-adjoint operator. The sum of these operators is the desired $\tilde{D}_X$. Here is the definition.

DEFINITION 6.7. $\tilde{D}_X = \sum_\alpha \rho_\alpha \tilde{D}_{X,\alpha} \rho_\alpha$.

3.4. Construction of (D_ν, h_ν) and Its Properties. Through the constructions in (a), (b), (c), D_ν is defined in Definition 6.5.

We study properties of $\tilde{D}_X, \tilde{D}_{\text{fiber}}$ and D_ν, which are necessary to obtain the index of D_ν.

Recall that we could not construct the lift $\tilde{D}_X$ of D_X so that it enjoys the "desired property" in full generality, which was the first candidate. Namely, $\tilde{D}_X(\pi^* a \otimes \tilde{b})$ and $\pi^*(D_X a) \otimes \tilde{b}$ are not necessarily equal to each other for an arbitrary section a of W_X and an arbitrary section $\tilde{b}$ of W_{fiber}. However, this equality holds for a special choice of $\tilde{b}$.

The way of choosing $\tilde{b}$ is as follows. $W_{\mathbb{R}^m}$ is a vector bundle on $\mathbb{R}^m$ and the $O(m)$-action on $\mathbb{R}^m$ naturally lifts to it. Pick an $O(m)$-invariant section $\bar{b}$ of this vector bundle. Then take a copy of $\bar{b}$ on each fiber on X and gather them to obtain a section of W_{fiber}.

LEMMA 6.8. *If $\tilde{b}$ is given by $\bar{b}$ as above, the following equality holds for any a:*

$$\tilde{D}_X(\pi^* a \otimes \tilde{b}) = \pi^*(D_X a) \otimes \tilde{b}\,.$$

PROOF. First of all, we consider the problem on $\nu|U_\alpha = U_\alpha \times \mathbb{R}^m$. The point is that $\tilde{b}$ is equal to $\pi_\alpha^* \bar{b}$. Using the product structure, we find that

$$\tilde{D}_{X,\alpha}(\pi^* a \otimes \pi_\alpha^* \bar{b}) = \pi^*(D_X a) \otimes \pi_\alpha^* \bar{b}.$$

Replace a by $\rho_\alpha a$ in the above equality, multiply ρ_α and take the sum over α. Then we obtain the desired relation, with D_X replaced by $D'_X = \sum_\alpha \rho_\alpha D_X \rho_\alpha$. Then we only need to check $D'_X = D_X$. The commutator $[\rho_\alpha, D_X]$ is an operator, which does not contain derivatives. Hence the twice commutator $[\rho_\alpha, [\rho_\alpha, D_X]]$ is 0. Expand this equality and take the sum over all α. We obtain $2D_X - 2D'_X = 0$. □

LEMMA 6.9. *The following commutative relations hold:*

(1) $\tilde{D}_X \tilde{D}_{\text{fiber}} = \tilde{D}_{\text{fiber}} \tilde{D}_X$.
(2) $\tilde{D}_X \tilde{h}_{\text{fiber}} = \tilde{h}_{\text{fiber}} \tilde{D}_X$, $\quad \tilde{h}_X \tilde{D}_{\text{fiber}} = \tilde{D}_{\text{fiber}} \tilde{h}_X$, $\tilde{h}_X \tilde{h}_{\text{fiber}} = \tilde{h}_{\text{fiber}} \tilde{h}_X$.

PROOF. We only show that $\tilde{D}_X \tilde{D}_{\text{fiber}} = \tilde{D}_{\text{fiber}} \tilde{D}_X$. Other commutative relations are shown in a similar way (easier).

$\tilde{D}_{\text{fiber}}$ commutes with ρ_α. (While the former only contains derivatives along fibers on X, the latter is constant after restriction to each fiber.)

$\tilde{D}_{\text{fiber}}$ commutes with $D_{X,\alpha}$ on $\nu|U_\alpha$. (The former contains only derivatives in the directions of the second factor, and the latter contains only derivatives in the directions of the first factor with respect to the product structure on $U_\alpha \times \mathbb{R}^m$.) Therefore $\tilde{D}_{\text{fiber}}$ commutes with $\tilde{D}_X$, which is expressed as a polynomial of such operators. □

As an operator on $\mathbb{R}^m$, the supersymmetric harmonic oscillator $(W_{\mathbb{R}^m}, D_{\mathbb{R}^m}, h_{\mathbb{R}^m})$ (with cylindrical end) practically means $D_{\mathbb{R}^m,t} = D_{\mathbb{R}^m} + th_{\mathbb{R}^m}$ for a sufficiently large $t > 0$.

Based on the preparation above, we study the index of $D_{\nu,t} = D_\nu + th_\nu$. From now on, we write $\tilde{D}_{X,t} = \tilde{D}_X + t\tilde{h}_X$, $\tilde{D}_{\text{fiber},t} = \tilde{D}_{\text{fiber}} + t\tilde{h}_{\text{fiber}}$, etc. Then the equality

$$D_{\nu,t} = \epsilon_{W_X} \circ \tilde{D}_{X,t} + \tilde{D}_{\text{fiber},t}$$

holds. By Lemma 6.9, $\tilde{D}_{X,t}$ and $\tilde{D}_{\text{fiber},t}$ commute with each other. Thus, in the above equality, the first term and the second term in the right hand side are anti-commutative. Hence we find that

$$D_{\nu,t}^2 = \tilde{D}_{X,t}^2 + \tilde{D}_{\text{fiber},t}^2.$$

Since $\tilde{D}_{X,t}$ and $\tilde{D}_{\text{fiber},t}$ are formally self-adjoint, the following lemma is intuitively plausible.

LEMMA 6.10.

(1) $\operatorname{Ker} \tilde{D}_{X,t}^2 = \operatorname{Ker} \tilde{D}_{X,t}$.

(2) $\operatorname{Ker} D_{\nu,t}^2 = \operatorname{Ker} \tilde{D}_{X,t}^2 \cap \operatorname{Ker} \tilde{D}_{\text{fiber},t}^2$.

PROOF. Since the arguments are similar, we show only the second statement. It is clear that the right hand side is contained in the left hand side.

For the time being, we take it for granted that, for an element s of the left hand side, both sides of the following equality converge, and the equality holds for an element s of the left hand side $\operatorname{Ker} D_{\nu,t}^2$:

$$\int_\nu (s, \tilde{D}_{X,t}^2 s) = \int_\nu |\tilde{D}_{X,t} s|^2, \tag{6.1}$$

$$\int_\nu (s, \tilde{D}_{\text{fiber},t}^2 s) = \int_\nu |\tilde{D}_{\text{fiber},t} s|^2. \tag{6.2}$$

This follows from a formal manipulation using the integration by parts. What we take for granted for the time being is that the contribution from the end is 0.

Adding these two equalities, the left hand side becomes 0. Since the right hand side is 0, we find that s belongs to $\operatorname{Ker} \tilde{D}_{X,t}$ and $\operatorname{Ker} \tilde{D}_{\text{fiber},t}$, which concludes the proof.

s (and its derivatives of any order) decays exponentially at the end of ν by the exponential decay and the a-priori estimate in Chapter 5. On the other hand, $\tilde{D}_X$, resp. $\tilde{D}_{\text{fiber}}$, which is the part of $\tilde{D}_{X,t}$, resp. $\tilde{D}_{\text{fiber},t}$, involving derivatives, satisfies Condition (D), hence the integration by parts as above is justified. (Details are left to readers.) □

Now we take an element f of $\operatorname{Ker} D^2_{\nu,t}$. By the above lemma, f turns out to be an element of $\operatorname{Ker} \tilde{D}^2_{\text{fiber},t}$. Note that $\tilde{D}_{\text{fiber},t}$ is defined by gathering $D_{\mathbb{R}^m,t}$ for each fiber on X.

Denote by $\tilde{b}_t$ the section of W_{fiber} constructed by gathering $O(m)$-invariant elements $\bar{b}_t$ in Lemma 4.15. Then, by Lemma 4.15, we find that there exists a unique section a of W_X such that f is expressed in the form $\pi^* a \otimes \tilde{b}_t$. Moreover, since the same element belongs to $\operatorname{Ker} \tilde{D}_{X,t}$,

$$0 = \tilde{D}_{X,t}(\pi^* a \otimes \tilde{b}_t) = \pi^*(D_{X,t} a) \otimes \tilde{b}_t$$

must hold by Lemma 6.8. This implies that a belongs to $\operatorname{Ker} D_{X,t}$.

Conversely, for an element a of $\operatorname{Ker} D_{X,t}$, it is clear by following the above argument in a reverse way that $\pi^* a \otimes \tilde{b}_t$ belongs to $\operatorname{Ker} D^2_{\nu,t}$.

In sum, we find the following.

LEMMA 6.11. $\operatorname{Ker} D^2_{\nu,t} \cong \operatorname{Ker} D^2_{X,t}$.

By the construction, the above isomorphism preserves the degree. Therefore the following result, which we aimed at, immediately follows from the definition of the index.

LEMMA 6.12. $\operatorname{ind} D_{\nu,t} = \operatorname{ind} D_{X,t} D$

3.5. D_ν as an Operator of Dirac Type. Next, we show that D_ν is regarded as an operator of Dirac type with an appropriate choice of a Riemannian metric on the total space of ν.

Now we explain the way of equipping the Riemannian metric on the total space of ν, which we postponed. (Up to now, each fiber on X is equipped with a Riemannian metric. But we did not equip the total space with a Riemannian metric.) When the total space of ν is equipped with a Riemannian metric, it is sufficient to check that the following conditions are satisfied, in order to regard D_ν as an operator of Dirac type.

(1) Existence of an isometric isomorphism between $T\nu$ and $\pi^* TX \oplus T_{\text{fiber}}\nu$.
(2) The principal symbol of D_ν fulfills the definition of operators of Dirac type with respect to the Riemannian metric.
(3) Moreover, as a technical assumption for the index to be defined, Conditions (R) and (D) are satisfied.

We study these conditions one by one.

In order for Condition (1) to be satisfied, it is sufficient to define the Riemannian metric on $T\nu$ by the Riemannian metric on $T_{\text{fiber}}\nu \oplus \pi^*TX$ through the isomorphism $T\nu \cong T_{\text{fiber}}\nu \oplus \pi^*TX$ obtained from a splitting of the natural exact sequence

$$0 \to T_{\text{fiber}}\nu \to T\nu \to \pi^*TX \to 0.$$

Namely, we put the Riemannian metric studied in Lemma 3.24 in Chapter 3. Hence it satisfies Condition (R).

In order that other conditions are satisfied, we choose a concrete splitting as follows. On $\nu|U_\alpha$, define the splitting using the product structure $U_\alpha \times \mathbb{R}^m$ in an obvious way. The splitting is expressed in terms of a left inverse θ_α of the natural map $\pi^*T^*X \to T^*\nu$ on $\nu|U_\alpha$. Here the glued $\theta = \sum_\alpha (\pi^*\rho^2)\theta_\alpha$ is a global left inverse of the natural map $\pi^*T^*X \to T^*\nu$, which gives a splitting everywhere on ν. Using this splitting, we equip ν with a Riemannian metric.

Next we would like to check that the principal symbol of D_ν fulfills the definition of operators of Dirac type as presented in Condition (2). The principal symbol of D_ν is

$$\sigma(D_\nu) = \epsilon_{W_X} \circ \sigma(\tilde{D}_{\text{fiber}}) + \sigma(\tilde{D}_X).$$

LEMMA 6.13.

$$\sigma(\tilde{D}_X): \quad T^*\nu \to \operatorname{End}(\pi^*W_X \otimes W_{\text{fiber}})$$

coincides with the composition of $\theta : T^*\nu \to \pi^*T^*X$ *and*

$$\pi^*\sigma(D_X) \otimes \operatorname{id}_{W_{\text{fiber}}} : \quad T^*X \to \operatorname{End}(\pi^*W_X) \otimes \operatorname{End}(W_{\text{fiber}}).$$

PROOF. Recall the way of our construction by gluing to find that

$$\sigma(\tilde{D}_X) = \sum_\alpha (\pi^*\rho_\alpha)^2 \sigma(\tilde{D}_{X,\alpha}).$$

Here, by the product structure $U_\alpha \times \mathbb{R}^m$ of $\nu|U_\alpha$, $\sigma(\tilde{D}_{X,\alpha})$ coincides with the composition of θ_α and $\pi^*\sigma(D_X) \otimes \operatorname{id}_{W_{\text{fiber}}}$. Hence we obtain the desired relation from $\theta = \sum_\alpha \rho_\alpha^2 \theta_\alpha$. □

Note that $\sigma(D_X)$ is equal to c_{W_X}. Then we can express the above coincidence by the identification $T^*\nu \cong \pi^*T^*X \oplus T^*_{\text{fiber}}\nu$ obtained from θ in the following way: the principal symbol of $\tilde{D}_X$ does not depend on the $T^*_{\text{fiber}}\nu$-component of $T^*\nu$ and is determined only by the π^*T^*X-component. Its value is given by

$$\sigma(\tilde{D}_X) = \pi^* c_{W_X} \otimes \operatorname{id}_{W_{\text{fiber}}}.$$

On the other hand, we immediately find the principal symbol of $\tilde{D}_{\text{fiber}}$, which does not depend on the $\pi^* T^* X$-component of $T^*\nu$ and is determined only by the $T^*_{\text{fiber}}\nu$-component. Its value is given by

$$\sigma(\tilde{D}_{\text{fiber}}) = \text{id}_{\pi^* W_X} \otimes c_{W_{\text{fiber}}}.$$

Thus, we find that

$$\begin{aligned} \sigma(D_\nu) &= \pi^* c_{W_X} \otimes \text{id}_{W_{\text{fiber}}} + \epsilon_{\pi^* W_X} \otimes c_{W_{\text{fiber}}} \\ &= c_{W_\nu}. \end{aligned}$$

Hence D_ν is an operator of Dirac type.

As for Condition (3), we can find that Condition (R) follows from Lemma 3.24 in Chapter 3. We can easily check Condition (D) by following our construction.

Therefore we showed that all three conditions are satisfied for the constructed Riemannian metric.

4. Invariance of the Index: Proof 2 (Excision Theorem)

We prove that $(W_\nu, [h_\nu])$ and $(W_E, [h_E])$ have the same index. This is an important step, though the essential part is already contained in the argument in Chapter 3, which shows that the index can be defined for operators on open manifolds.

Firstly, if a pair on U is the restriction of a pair on E, the Excision Theorem (Theorem 3.29 in Chapter 3) guarantees that their index takes the same value both on U and on E. (This is an important point!).

Thus, it suffices to study the index on U or ν. By the way of our construction, we find that

$$\text{ind}(W_E, [h_E]) = \text{ind}(W_\nu, [h_\nu]) + \text{ind}(F' \otimes (S_E|U), [h' \otimes \text{id}_{S_E}|U]).$$

Note that the support of h' is empty, hence the second term of the right hand side is 0.

5. Pairs on Even Dimensional Euclidean Spaces

5.1. Construction 3 (Construction of the Element in the Homotopy Group of U). Let $(W_E, [h_E])$ be a pair on the Euclidean space E of dimension $2N$. Fix an orientation of E. Then it determines a unique spin structure. Denote by S_E the corresponding spinor bundle. Then there exist a $\mathbb{Z}_2$-graded Hermitian vector bundle

F_E on E and a Hermitian endomorphism h_F of F_E of degree 1 such that

$$W_E = F_E \otimes S_E, \quad h_E = h_{F_E} \otimes \mathrm{id}_{S_E}.$$

By perturbing h_E by homotopy, we assume that it is unitary and Hermitian automorphism outside of a compact region K. Since E is contractible, there exists a trivialization

$$F_E \cong E \times (\mathbb{C}^r \oplus \mathbb{C}^r).$$

On the sphere S^{2N-1}, which surrounds K in its inner domain, h_E is expressed as

$$h_E = \begin{pmatrix} 0 & f^* \\ f & 0 \end{pmatrix}$$

using some continuous map $f : S^{2N-1} \to U(r)$. Here, the homotopy class of f depends only on the pair $(W_E, [h_E])$. Denote also by f the composition of f and the embedding $U(r) \to U$. The desired definition of the element of $\pi_{2N-1}(U)$ is given by the homotopy class $[f]$.

5.2. Proof 3 (The Index is Determined by $[f]$). In general, suppose that an element f of $\pi_{2N-1}(U)$ is given. By the definition of the topology on U, the image of f is contained in $U(r)$ for a sufficiently large r. Then we define h_F by the above formula outside of S^{2N-1} and extend it as a Hermitian endomorphism of degree 1 to the inside of S^{2N-1} to construct a pair $(W_E, [h_E])$. It suffices to show that the index of this pair is determined only by $[f]$.

Firstly, for different r's, we obtain two pairs. We show that the indices of them are the same. In fact, take a larger r_1 in place of r. The newly obtained pair is the direct sum of the pair constructed with r and a pair with empty support. Since the index of a pair with empty support is 0, the direct sum with it does not change the index.

Secondly, we show that the indices of f_0 and f_1 are the same, if f_0 and f_1 are joined by a homotopy. In fact, by the definition of the topology on U, the image of the homotopy is totally contained in $U(r)$ for a sufficiently large r. With this r, we obtain a homotopy of pairs, which implies the coincidence of the indices.

Therefore we have proved that the index depends only on $[f]$.

SUMMARY.

6.1 For a given $\mathbb{Z}_2$-graded Clifford module $(W_X, [h_X])$ on X with compact support, embed X into the Euclidean space E of a sufficiently large even dimension. Then we can construct the $\mathbb{Z}_2$-graded Clifford module $(W_E, [h_E])$ on E with compact support.

6.2 The first point of the construction is as follows. Construct a family of supersymmetric harmonic oscillators associated with the normal bundle of X and then take the "product" with it.

6.3 The second point of the construction is to extend the "product" to the outside of a tubular neighborhood of X in such a way that the Excision Theorem can be applied.

6.4 The indices of $(W_E, [h_E])$ and $(W_X, [h_X])$ coincide. To show it, we use the fact that the index of the supersymmetric harmonic oscillator is 1 and the multiplicativity of the index.

6.5 $\mathbb{Z}_2$-graded Clifford modules on X with compact support are classified by elements of the homotopy group of the unitary group. By this, the computation of the index reduces to the computation in the homotopy group of U.

CHAPTER 7

Characteristic Classes

We introduce characteristic classes as terminology to describe how the vector bundle is twisted. In this book, we explain Chern-Weil theory, which produces them, in the case that the base spaces are manifolds, as elements of the de Rham cohomology using connections and their curvatures of vector bundles.

Firstly, we introduce the Chern character, which provides a fundamental relation between vector bundles and de Rham cohomology classes. The relative Chern character is defined for $\mathbb{Z}_2$-graded vector bundles. At this point, a kind of "localization" plays a key role.

Next we prove the Thom isomorphism theorem for the de Rham cohomology. It is a prototype of the Thom isomorphism theorem (containing the Bott periodicity theorem) for vector bundles (K-groups), which we will discuss in Chapter 9 of the sequel. We also define the Euler class using the Thom class.

1. Connection and Curvature

The exterior derivative d is an operator, which acts on sections of $\bigwedge^{\bullet} T^*X$. d enjoys the following two fundamental properties. Firstly, d^2 is identically 0, although it is, at first glance, the second order differential operator, since d is a first order differential operator. Secondly, there holds a formula for the exterior derivative of the wedge product, namely, the following two properties:

$$d^2 = 0, \quad d(\omega_0 \wedge \omega_1) = (d\omega_0) \wedge \omega_1 + (-1)^{\deg \omega_0} \omega_0 \wedge d\omega_1.$$

The principal symbol of d, i.e.,

$$\sigma_1(d) : \ T^*X \to \mathrm{End}\left(\bigwedge^{\bullet} T^*X\right),$$

is the map by the wedge product from the left

$$v \mapsto v^{\wedge}.$$

Let F be a complex vector bundle on a manifold X of rank r. If F is a trivial vector bundle $X \times \mathbb{C}^r$, a natural extension of d is defined on $\Omega^\bullet(F) = \Gamma((\bigwedge^\bullet T^*X \otimes \mathbb{C}) \otimes F)$. In fact, denote by $d^{(r)}$ the ordered r copies of d. Then

$$d^{(r)} : \Omega^\bullet(X \times \mathbb{C}^r) \to \Omega^\bullet(\mathrm{X} \times \mathbb{C}^{\mathrm{r}})$$

enjoys the above two properties. Here, in the latter product formula, ω_0 is considered as a differential form and ω_1 is considered as an element of $\Omega^\bullet(F)$.

Then, if F is not trivial, is there any extension of d in some sense?

Suppose that X is covered by a locally finite family of open subsets $\{U_\alpha\}$ such that the restriction of F to each U_α is isomorphic to the trivial bundle $U_\alpha \times \mathbb{C}^r$. Fix such an isomorphism for each α. Then, on each U_α, an extension $d_\alpha^{(r)}$ of the exterior derivative is defined and enjoys the two properties. As an attempt, set the operator

$$d_A : \Omega^\bullet(F) \to \Omega^\bullet(F) \tag{7.1}$$

by the weighted mean $d_A s = \sum_\alpha \rho_\alpha d_\alpha^{(r)} \rho_\alpha s$, where $\{\rho_\alpha^2\}$ is a partition of unity. (It is a similar construction to Lemma 2.44.) d_A is a first order differential operator raising the degree by 1. d_A satisfies

$$d_A(\omega \wedge s) = (d\omega) \wedge s + (-1)^{\deg \omega} \omega \wedge d_A s \tag{7.2}$$

for a differential form ω and an element s of $\Omega^\bullet(F)$. (It is obtained by summing them up after multiplying ρ_α to the product formula for $d_\alpha^{(r)}$ and $\omega \wedge \rho_\alpha s$ on U_α.)

d_A^2, however, is not zero any more. This is, first, because ρ_α, used from the partion of unity, and its derivatives appear. Second, on the intersection of distinct open subsets $U_\alpha, U_{\alpha'}$, $d_\alpha^{(r)}$ and $d_{\alpha'}^{(r)}$ are different operators, in general. (Therefore the clutching function and its derivatives appear.)

Since d_A is a first order differential operator, d_A^2 is a-priori a second order differential operator. But, similar cancellation occurs as in the case of the exterior derivative; hence we find that d_A^2 is an operator, which does not involve derivatives.

The point is that

$$\begin{aligned} d_A^2(\omega \wedge s) &= d_A((d\omega) \wedge s + (-1)^{\deg \omega} \omega \wedge d_A s) \\ &= d^2 \omega \wedge s + (-1)^{\deg d\omega} d\omega \wedge d_A s \\ &\quad + (-1)^{\deg \omega} d\omega \wedge d_A s + \omega \wedge d_A^2 s \\ &= \omega \wedge d_A^2 s \end{aligned}$$

holds for a differential form ω. This implies (by taking a local trivialization and studying the operator locally) that there exists a certain unique element F_A of $\Omega^2(\operatorname{End} F)$ such that

$$d_A^2 s = F_A \wedge s$$

holds. Here the product on the right hand side is the operation of applying an element of $\operatorname{End} F$ on F as well as taking the wedge product. We leave readers to check the details.

Based on the above discussion, we give the following definition.

DEFINITION 7.1. If a linear differential operator $d_A : \Omega^\bullet(F) \to \Omega^\bullet(F)$ raises the degree by 1 and enjoys property (7.2), we call d_A a **connection** on F.

As we have seen above, there exist connections for any F. By the way of the construction, you may realize that there are infinitely many connections on F. In fact, we easily see that the following lemma holds.

LEMMA 7.2.

(1) *If a first order differential operator $d_A : \Omega^0(F) \to \Omega^1(F)$ satisfies $d_A(fs) = (df) \wedge s + f \wedge d_A s$ for a function f and an element s of $\Omega^0(F)$, we can extend it uniquely in such a way that property (7.2) holds. Hence it gives a connection on F.*

(2) *The necessary and sufficient condition for a first order differential operator $d_A : \Omega^0(F) \to \Omega^1(F)$ to satisfy the above condition is that its principal symbol is expressed as*

$$T^*X \to \operatorname{Hom}(F, (T^*X \otimes \mathbb{C}) \otimes F), \quad v \mapsto (e \mapsto v \otimes e).$$

(3) *Suppose that d_A is a connection on F. For any element a of $\Omega^1(\operatorname{End} F)$, $d_A + a$ is also a connection. Moreover, any connection on F is written uniquely in this form.*

Using (1), (2) in the above lemma, we can define a connection by the principal symbol. Then the existence of a connection follows from Lemma 5.29 in Chapter 5. The above construction of a connection by gluing is just a direct argument following the proof of Lemma 5.29, Chapter 5, in this case.

For any connection d_A, d_A^2 becomes an operator, which does not contain derivatives.

DEFINITION 7.3. There is a unique element F_A of $\Omega^2(\operatorname{End} F)$ such that

$$d_A^2 s = F_A \wedge s.$$

We call this F_A the **curvature** of the connection d_A. (Note: The notation F_A for the curvature may be confused with the notation F for a vector bundle. When we use F as the notation for the curvature, we always put the suffix indicating the connection.)

The curvature F_A is one of the indicators which measure how the connection d_A is remote from $d^{(r)}$ defined on the trivial vector bundle. (Although F_A is identically 0, F is not necessarily isomorphic to a trivial vector bundle.)

REMARK 7.4. We give some explanation of the notation. There are other equivalent definitions of connections than the one we adopted in this book. Briefly, there are

- a first order differential operator $d_A : \Omega^0(F) \to \Omega^1(F)$, which enjoys some property (cf. Lemma 7.2 (2)),
- a differential 1-form on the principal $GL_r(\mathbb{C})$-bundle associated with F, which enjoys some property (called **principal connection**),
- a way of decomposing the tangent space of the total space of F into the direct sum of the part along fibers and its complement, which enjoys some property,

and so forth. When one of these definitions is adopted, the first order differential operator $d_A : \Omega^\bullet(F) \to \Omega^1(F)$ of degree 1 is often called the **covariant derivative** (associated with the principal connection or the connection in terms of the direct sum decomposition). In this book, we do not use these definitions in a direct way. But, as a way of using the notation, the symbol A, for example, stands for the geometric object indicated by these definitions above. The covariant derivative and the curvature which come from it are denoted by the symbols d_A, F_A, respectively. In this sense, we sometimes use the terminology "connection A".

LEMMA 7.5. *Let d_A be a connection on a vector bundle F on a manifold X. For another manifold Y and an arbitrary smooth map $\phi : Y \to X$, the connection $d_{\phi^* A}$ on the pull-back bundle $\phi^* F$ can be uniquely defined such that the following formula.*

$$d_{\phi^* A}\phi^* s = \phi^* d_A s.$$

Here, s is an arbitrary section of F.

Considering the issue locally, the proof is easy. It is because, in the case that F is trivial and $d_A = d^{(r)} + a$, ϕ^*F is also trivial, and we can set $d_{\phi^*A} = d^{(r)} + \phi^* a$ for our purpose.

Associated with a connection d_A on F, there exists a unique connection on $\operatorname{End} F$ with the following property. Let s be an element of $\Omega^\bullet(F)$ and f an element of $\Omega^\bullet(\operatorname{End} F)$. Then the following equality between elements of $\Omega^\bullet(\operatorname{End} F)$ holds:

$$d_A(f \wedge s) = (d_A f) \wedge s + (-1)^{\deg f} f \wedge d_A s.$$

Here, denote by the same symbol d_A the operator, which should be the connection on $\Omega(\operatorname{End} F)$. (Check that the operator $d_A(f\wedge) - (-1)^{\deg f}(f\wedge)d_A$, which assigns s to $d_A(f \wedge s) - (-1)^{\deg f} f \wedge d_A s$, does not contain derivatives. Then show that it is written by the action of some element of $\Omega^\bullet(\operatorname{End} F)$.)

Using this notation, the following Bianchi identity holds.

LEMMA 7.6 (Bianchi identity). $d_A F_A = 0$.

PROOF. It follows from the fact that $d_A^3 = (F_A\wedge)d_A = d_A(F_A\wedge)$ as operators on $\Omega^\bullet(F)$. □

2. Chern Character and Chern Classes

Note the following lemma in order to relate the connection as a twisted exterior derivative and the de Rham complex, which comes with the genuine exterior derivative.

LEMMA 7.7. *For* $\operatorname{trace}\colon \operatorname{End} F \to X \times \mathbb{C}$, *the following equality holds as maps from* $\Omega^\bullet(\operatorname{End} F)$ *to* $\Omega^\bullet(X \times \mathbb{C})$*:*

$$d \cdot \operatorname{trace} = \operatorname{trace} \cdot\, d_A.$$

PROOF. It suffices to show it locally. It is clear in the case that $d_A = d^{(r)}$. Note that the trace of the commutator of two matrices is 0 to find that it also holds for $d_A + a$ clearly. □

Then, by the Bianchi identity $d_A F_A = 0$, we immediately obtain that $d(\operatorname{trace} F_A) = 0$. Namely, $\operatorname{trace} F_A$ is a closed differential form, which defines a de Rham cohomology class of degree 2.

There are infinitely many choices of connections A on F, according to which the curvature F_A can change in various ways. However, we can find that the de Rham cohomology classes defined by $\operatorname{trace} F_A$ are the same as follows. For two given connections d_{A_0}, d_{A_1}, pick a homotopy d_{A_t} joining them so that we may assume that it extends smoothly to $t \in \mathbb{R}$ (for example, pick $d_{A_t} = (1-t)d_{A_0} + td_{A_1}$). Define the connection $d_{\tilde{A}}$ on the vector bundle $\mathbb{R} \times F$ on $\mathbb{R} \times X$ by

$$d_{\tilde{A}} s = d_{A_t} s + dt \wedge \partial_t s.$$

Then $\operatorname{trace} F_{\tilde{A}}$ derived from the curvature $F_{\tilde{A}}$ is a closed differential form of degree 2, which defines a de Rham cohomology class on $\mathbb{R} \times X$. Its restriction to $\{0\} \times X$, resp. $\{1\} \times X$, is $\operatorname{trace} F_{A_0}$, resp. $\operatorname{trace} F_{A_1}$. This implies that the de Rham cohomology classes defined by them are equal.

By a similar argument, for any natural number k, $\operatorname{trace}(F_A)^k$ is a closed differential form of degree $2k$ and defines a de Rham cohomology class, which does not depend on the choice of a connection A but only depends on the vector bundle F.

DEFINITION 7.8. The **Chern character** $chF = \sum_k ch_k F$ of a complex vector bundle F is defined to be the de Rham cohomology class by

$$\begin{aligned} chF &= \left[\operatorname{trace} \exp \frac{\sqrt{-1}\, F_A}{2\pi}\right] \in H^{\bullet}(X), \\ ch_k F &= \frac{1}{k!} \left[\operatorname{trace} \left(\frac{\sqrt{-1}\, F_A}{2\pi}\right)^k \in H^{2k}(X)\right], \end{aligned}$$

which does not depend on the choice of a connection A.

We examine the fact that the Chern character is independent of the choice of a connection A in a bit more concrete manner. From now on, we set

$$\begin{aligned} \tilde{F}_A &= \frac{\sqrt{-1}\, F_A}{2\pi}, \\ ch(A) &= \operatorname{trace} \exp \tilde{F}_A. \end{aligned}$$

Let A_t be a family of connections parametrized by the real number t and define a connection $\tilde{A}$ on the vector bundle $\tilde{A}$ on $\mathbb{R} \times X$ by the same formula as in the above proof. Decompose the closed differential form $ch(\tilde{A})$ on $\mathbb{R} \times X$ into the component containing dt and the component which does not contain it and write it as $dt \wedge \phi + \psi$. Then the restriction of ψ to $\{t\} \times X$ coincides with $ch(A_t)$. Since $ch(\tilde{A})$ is a closed differential form, $d_X \phi = \partial_t \psi$ holds. d_X appearing in the first term is the exterior derivative only in the X-directions. Integrate it in t from t_0 to t_1 to obtain

$$d_X \int_{t_0}^{t_1} \phi dt = ch(A_{t_1}) - ch(A_{t_0}).$$

This implies that the de Rham cohomology class $ch(A_t)$ is independent of t.

By the identification $H^0(X) = \mathbb{R}$, it follows from the definition that $ch_0(F)$ coincides with the rank of F. In the definition of Chern classes, we multiplied the factor $\sqrt{-1}/2\pi$ to the curvature in order to make the following normalization.

LEMMA 7.9. *Let L be a complex line bundle on a closed Riemann surface M. Then*

$$\deg L = \int_M ch_1 L$$

holds. Here, M is equipped with the orientation determined by the complex structure.

QUESTION 1. Prove the above lemma. (Hint: fix a section s of L, which intersects with the zero section transversely. Then $\deg L$ counts the number of zeros of s with signs. Here, L can be constructed by twisting the trivial complex line bundle $M \times \mathbb{C}$ in neighborhoods of zeros of s. Pick a connection A, which coincides with the exterior derivative outside of neighborhoods, where we make the twisting. Compute $ch_1 L$ using this connection.)

Fundamental properties of the Chern character are the following additivity and multiplicativity.

LEMMA 7.10. *Let F, F' be complex vector bundles on X:*

$$\begin{aligned} ch(F \oplus F') &= chF + chF', \\ ch(F \otimes F') &= chF chF'. \end{aligned}$$

PROOF. We only explain the multiplicativity. (The additivity is similar.) Let A, A' be connections on F, F', respectively. We define a connection $A \otimes A'$ on $F \otimes F'$ by

$$d_{A\otimes A'}(s \otimes s') = (d_A s) \otimes s' + (-1)^{\deg s} s \otimes (d_{A'} s').$$

Then

$$\tilde{F}_{A\otimes A'} = \tilde{F}_A \otimes \mathrm{id}_{F'} + \mathrm{id}_F \otimes \tilde{F}_{A'}$$

holds as elements of $\Omega^2(\mathrm{End}(F \otimes F')) = \Omega^2(\mathrm{End}\, F \otimes \mathrm{End}\, F')$. Take exp of it. Then the summation on the right hand side becomes the tensor product. Apply trace to it, then it becomes just the product (exterior product). This yields the desired multiplicativity. □

We define the Chern classes $c(F) = \sum_k c_k(F)$, which are defined in an almost parallel way to the Chern character and give essentially equivalent information to the Chern character, as follows.

DEFINITION 7.11. For a connection A on a complex vector bundle F, we define a differential form $c(A)$ by

$$c(A) = \det(\mathrm{id}_F + \tilde{F}_A).$$

$c(A)$ consists only of even degree components. Denote by $c_k(A)$ the component of degree $2k$. Then $c(A)$ resp. $c_k(A)$ are closed differential forms and the de Rham cohomology class $[c(A)]$, resp. $[c_k(A)]$ is independent of the choice of A and determined only by F. We call the de Rham cohomology class the **Chern class**, resp. the k-th **Chern class** and denote it by $c(F)$, resp. $c_k(F)$.

By the definition, we find that $c_0(F) = 1$ and $c_k(F) = 0$, if k is larger than the rank of F (as a complex vector bundle).

The components of the Chern character and the Chern classes are expressed in polynomials of each other. In fact, the closed differential forms $ch(A)$ and $c(A)$, which appear in the step before taking the de Rham cohomology classes, are expressed in polynomials of each other as follows.

Let R be a commutative algebra over $\mathbb{C}$ with the unit, in general. $R \otimes \mathrm{End}(\mathbb{C}^r)$ is linearly isomorphic to the direct sum of r^2 copies of R. Denote by Φ_{ch_k}, Φ_c the two maps from $R \otimes \mathrm{End}(\mathbb{C}^r)$ to R

$$\Phi_{ch_k} a = \mathrm{trace}\, \frac{1}{k!} a^k, \quad \Phi_c a = \det(I_r + a),$$

respectively. I_r is the unit matrix of size r. Both Φ_{ch_k} and Φ_c are polynomials of r^2 elements of R with rational coefficients and

$$\Phi_{ch_k}(g^{-1}ag) = \Phi_{ch_k} a, \quad \Phi_c(g^{-1}ag) = \Phi_c a$$

holds for an element g of $GL(\mathbb{C}^r)$.

Fix a trivialization $(F)_x \cong \mathbb{C}^r$ of F at each point x of X to find that

$$ch_k(A)_x = \Phi_{ch_k}((\tilde{F}_A)_x), \quad c(A)_x = \Phi_c((\tilde{F}_A)_x)$$

holds in $R = \bigoplus_k \bigwedge^{2k}(T^*X)_x$. Hence, in order to find a relation between $ch_k(A)$ and $c(A)$, it suffices to find a relation between the polynomials Φ_{ch_k} and Φ_c with rational coefficients. This is purely a proposition in algebra. For our purpose, it is enough to work with the case of $R = \mathbb{C}$.

Let us introduce the following notation.

DEFINITION 7.12. Denote by R_r the commutative algebra consisting of maps Φ from $\mathrm{End}(\mathbb{C}^r)$ to $\mathbb{C}$, which is expressed as polynomials of the entries with complex coefficients, such that $\Phi(gag^{-1}) = \Phi a$ for an element of $GL(\mathbb{C}^r)$.

Diagonalizable elements are dense in $\mathrm{End}(\mathbb{C}^r)$. Thus an element of R_r is determined by the values at the diagonal matrices $\mathrm{diag}(x_1, x_2, \ldots, x_r)$. In fact, the following algebraic lemma is an exercise in linear algebra.

LEMMA 7.13. *R_r is isomorphic to the commutative algebra consisting of polynomials of $x_1, x_2, \ldots, x_r$ with complex coefficients, which are symmetric in $x_1, x_2, \ldots, x_r$.*

Using the isomorphism in this lemma, we find that

$$\Phi_{ch_k} = \sum_m \frac{1}{k!} x_m^k, \quad \Phi_c = \prod_m (1 + x_m).$$

Using the fact that R_r is generated, as an algebra, by elementary symmetric polynomials, the following lemma follows from this.

LEMMA 7.14. *Let F be a complex vector bundle on X of rank r. Then there exists a homomorphism as $\mathbb{R}$-algebras from R_r to $H^\bullet(X)$ such that the following property holds for each $0 < k \leq r$:*

- *The elementary symmetric polynomial $x_1 x_2 \cdots x_k + \cdots$ of $x_1, x_2, \ldots, x_r$ of degree k is mapped to $c_k(F)$.*
- *$\sum_m x_m^k / k!$ is mapped to $ch_k F$.*

REMARK 7.15.

(1) In general, characteristic classes of vector bundles are the rules which assign a vector bundle F to an element $\alpha(F)$ of the cohomology of the base space X, which is natural in the following sense. Namely, for F is a vector bundle on X and a continuous map $\phi : X \to Y$, the element $\alpha(\phi^* F)$ corresponding to the pull-back $\phi^* F$ coincides with the pull-back $\phi^* \alpha(F)$ of $\alpha(F)$ by ϕ.

(2) If we work with base spaces, which are quite general, the construction of characteristic classes gets technically difficult. In this book, we only consider the case that X is a

manifold and maps are smooth maps between manifolds. In this range, the Chern character and the Chern classes, which are defined above, enjoy the naturality with respect to the pull-back operation.

(3) It is known that Chern classes lift naturally to elements of cohomology with integer coefficients. But it is not necessarily the case for the Chern character.

(4) Using the existence of the classifying space, elements of the cohomology of the classifying space are known to correspond in a one-to-one way to characteristic classes.

(5) There exists a model of the classifying space using the **superspace.**[1] To give a map (as superspaces) to this model and to give a connection are almost verbatim. We can interpret the definition of the Chern character and the Chern classes in this way.

3. Localization of Chern Character

3.1. The Case of $\mathbb{Z}_2$-graded Vector Bundles. Let $F = F^0 \oplus F^1$ be a $\mathbb{Z}_2$-graded vector bundle on X. We define the Chern character of the $\mathbb{Z}_2$-graded vector bundle by

$$chF = chF^0 - chF^1.$$

Namely, pick a connection A on F, which is the direct sum of an arbitrary connection A^0 on F^0 and an arbitrary connection A^1 on F^1, to obtain

$$chF = \left[\operatorname{trace} \epsilon_F \exp \tilde{F}_A\right].$$

(The condition on A can be phrased that d_A is commutative with ϵ_F.)

We explain an extension of the above expression when an endomorphism h of F of degree 1 is given as well as a connection $A = A^0 \oplus A^1$ on F.

We prepare a little notation.

Denote by $\epsilon_\wedge$, resp. ϵ_F, the multiplication by ± 1 according to the parity of the degree of the exterior product, resp. the degree of F. Define the $\mathbb{Z}_2$-degree on $\Omega^\bullet(F)$ using the sum of the degree of the exterior product and the degree of F. The parity of this degree

[1] One of definitions of the "superspace" is a "manifold equipped with a sheaf of algebras over the sheaf of germs of smooth functions, which is locally isomorphic to the sheaf of germs of functions with values in $\bigwedge^\bullet \mathbb{R}^r$ for some r".

is determined accordingly so that the eigenvalue of $\epsilon_\wedge \epsilon_F$ is ± 1. We define the degree on $\Omega^\bullet(\operatorname{End} F)$ in a similar way. Denote by deg the degree of $\Omega^\bullet(F)$, $\Omega^\bullet(\operatorname{End} F)$. Both d_A and h are maps of degree 1.

For an element ω, ω' of $\Omega^\bullet(X)$, a section f of $\operatorname{End} F$ and a section s of F, define the product

$$\Omega^\bullet(\operatorname{End} F) \times \Omega^\bullet(F) \to \Omega^\bullet(F)$$

by $(\omega \otimes f,\ \omega' \otimes s) \mapsto (-1)^{\deg f \deg \omega'}(\omega \wedge \omega') \otimes fs$. Using this product, we can define the product

$$\{\,\cdot\,,\,\cdot\,\}:\ \Omega^\bullet(\operatorname{End} F) \times \Omega^\bullet(\operatorname{End} F) \to \Omega^\bullet(\operatorname{End} F)$$

in such a way that

$$f(gs) = (-1)^{\deg f \deg g} g(fs) + \{f, g\}s$$

holds for elements f, g of $\Omega^\bullet(\operatorname{End} F)$ and an element s of $\Omega^\bullet(F)$.

As an extension of the notion of the connection, we consider a map $d_{A,h} = d_A + (-1)^{1/4}h$ of degree 1. Here, $(-1)^{1/4} = \exp(\pi\sqrt{-1}/2)$. We multiplied $(-1)^{1/4}$ to h for a later purpose, and it is not essential at this moment. For an element ω of $\Omega^\bullet(X)$ and an element s of $\Omega^\bullet(F)$, we have

$$d_{A,h}(\omega s) = (d\omega)s + (-1)^{\deg \omega}\omega d_{A,h}s.$$

More generally, define, for an element f of $\Omega^\bullet(\operatorname{End} F)$,

$$d_{A,h}f = d_A f + (-1)^{\frac{1}{4}}\{h, f\}$$

to obtain

$$\begin{aligned} d_{A,h}(fs) &= (d_{A,h}f)s + (-1)^{\deg f} f d_{A,h}s \\ d_{A,h}\{f, g\} &= \{d_{A,h}f, g\} + (-1)^{\deg f}\{f, d_{A,h}g\}. \end{aligned}$$

REMARK 7.16. $d_{A,h}$ is a special case of the **superconnection** introduced by Quillen.

The square of $d_{A,h}$ is represented as the multiplication by an element $F_{A,h}$ of $\Omega^\bullet(\operatorname{End} F)$. In fact,

$$\begin{aligned} (d_A + (-1)^{\frac{1}{4}}h)^2 &= d_A^2 + (-1)^{\frac{1}{4}}(d_A \cdot h + h \cdot d_A) + \sqrt{-1}\, h^2 \\ &= F_A + (-1)^{\frac{1}{4}}(d_A h) + \sqrt{-1}\, h^2. \end{aligned}$$

Denote by $F_{A,h}$ and also by $\tilde{F}_{A,h}$ the multiple of $F_{A,h}$ by $\sqrt{-1}/2\pi$.

In a parallel way to the argument in the previous section, we find that

$$d_{A,h}\tilde{F}_{A,h} = 0.$$

Using the formula

$$\operatorname{trace} \epsilon_F \{f, g\} = 0$$

for elements f, g of $\Omega^\bullet(\operatorname{End} F)$, we easily see that

$$\operatorname{trace} \epsilon_F d_{A,h} = d \operatorname{trace} \epsilon_F \colon \ \Omega^\bullet(\operatorname{End} F) \to \Omega^\bullet(X) \otimes \mathbb{C}.$$

From this, we find that

$$ch(A, h) = \operatorname{trace} \epsilon_F \exp \tilde{F}_{A,h}$$

is a closed differential form and its de Rham cohomology class does not depend on the choice of A, h by an argument parallel to the one in the previous section. Here the difference from the previous section, where h is not present, is that exp in the definition of $ch(A)$ is eventually a polynomial, while exp in the definition of $ch(A, h)$ is an infinite series (of the matrix with entries being differential forms) defined by the Taylor expansion.

In particular, in the case that $h = 0$, this closed differential form coincides with the definition of the Chern character of the $\mathbb{Z}_2$-graded vector bundle F given in the beginning of this section. Hence, we obtain

LEMMA 7.17. *For a connection d_A, which commutes with ϵ_F for a complex vector bundle F, and an endomorphism h of degree* 1, *$chF = [ch(A, h)]$ holds.*

3.2. Relative Chern character. The merit of introducing not only the connection A but also h is that we can use them to define the Chern character for the $\mathbb{Z}_2$-graded vector bundle as an element of the relative de Rham cohomology.

DEFINITION 7.18. Let C be a closed subset of X. We define the **relative de Rham cohomology** $H^\bullet(X, C)$ as the cohomology of the complex consisting of all differential forms with support disjoining from C and the exterior derivative.

The following method of providing an element of the relative de Rham cohomology is useful. Let U be an open set containing C, ω a closed differential form on X, η a differential form on U such that $\omega = d\eta$ on U. For any function ρ which is equal to 1 on an open neighborhood of C, $\omega - d(\rho\eta)$ is a closed differential form with support disjoint from C. Then, as we easily see, the element of $H^\bullet(X, C)$ defined by $\omega - d(\rho\eta)$ does not depend on the choice of ρ and is determined only by ω and η.

DEFINITION 7.19. Denote by $[\omega, \eta]$ the element of $H^\bullet(X, C)$ defined by $\omega - d(\rho\eta)$.

Firstly, we explain an easy case of the definition of the relative Chern character.

Let C be a closed set of X and suppose that h satisfies that $h^2 = \mathrm{id}_F$ on some open neighborhood of C. Namely, write

$$h = \begin{pmatrix} 0 & h_{01} \\ h_{10} & 0 \end{pmatrix}.$$

Then h_{10} is an isomorphism from F^0 to F^1 on a neighborhood of C and h_{01} is the inverse of h_{01} at each point of C.

Pick any connections d_{A^0}, resp. d_{A^1}, on F^0, resp. F^1, which are identified on an open neighborhood of C by an isomorphism h_{10} and write $A = A_0 \oplus A_1$. Then $d_A h = 0$ holds on the open neighborhood of C.

Now we define the closed differential form $ch(A, h)$ by

$$ch(A, h) = \mathrm{trace}\, \epsilon_F \exp \tilde{F}_A = \mathrm{trace} \exp \tilde{F}_{A_0} - \mathrm{trace} \exp \tilde{F}_{A_1}, \tag{7.3}$$

then $ch(A, h)$ is identically 0 on the open neighborhood of C. Hence it defines an element of the relative de Rham cohomology $H^\bullet(X, C)$. We can show that this element does not depend on the choice of A^0, A^1 in a similar way to the previous section. As for information of h, we use only the restriction of h to the open neighborhood of C. This element is also invariant under continuous deformations of the endomorphism h of degree 1, keeping the condition that $h^2 = \mathrm{id}_F$ on some open neighborhood of C.

DEFINITION 7.20. Let h be an endomorphism of F of degree 1 defined on some neighborhood of C such that $h^2 = \mathrm{id}_F$. Denote the element of $H^\bullet(X, C)$ defined by (7.3) by $ch(F, [h])$ and call it the **relative Chern character**. $[h]$ is the homotopy class of h such that $h^2 = \mathrm{id}_F$ on some neighborhood of C.

REMARK 7.21. If the complement of C is relatively compact, there exists a natural map from $H^\bullet(X, C)$ to $H_c^\bullet(X)$. Hence, in this case, $ch(F, [h])$ gives an element of $H_c^\bullet(X)$. We also denote it by the same symbol $ch(F, [h])$.

The relative Chern character also enjoys the additivity formulated as follows. Let (F, C, h) be an object as above. Suppose also that we have (F', C', h') on X, similarly. We consider $(F \oplus F', C \cap C', h \oplus h')$. Here $h \oplus h'$ is defined only on some open neighborhood of $C \cap C'$.

On this open neighborhood, $(h \oplus h')^2 = \mathrm{id}_{F \oplus F'}$ holds, hence defines an element $ch(F \oplus F', [h \oplus h'])$ of $H^\bullet(X, C \cap C)$. Then the following property holds clearly.

LEMMA 7.22. $ch(F \oplus F', [h \oplus h']) = ch(F, [h]) + ch(F', [h'])$. *Here, we regard each term on the right hand side as an element of* $H^\bullet(X, C \cap C')$.

3.3. Definition of Relative Chern Character by Localization. We can extend the definition of the relative Chern character to the case that h does not necessarily satisfy $h^2 = \mathrm{id}_F$ on a neighborhood of C.

Let h be an endomorphism of F of degree 1 defined on X and C a closed subset of X. Suppose the following assumption.

Assumption (C, h)**.** h is an endomorphism of F of degree 1 and the real part of each eigenvalue of $(h^2)_x$ is positive for each point x in some open neighborhood of C.

In particular, h is an automorphism of F on the open neighborhood of C. By deforming h, and continuously keeping Assumption (C, h) satisfied, we can assume that $h^2 = \mathrm{id}_F$ holds on some open neighborhood of C (use the generalized eigenspace decomposition at each point, then join the component of h to $+id$ or $-id$ by the linear combination on each generalized eigenspace according to positivity or negativity of the real part of the generalized eigenvalue). Extend the previous definition to define $ch(F, [h])$ for the homotopy class of h satisfying Assumption (C, h).

For a real parameter t, we consider $d_{A,th}$. Recall that, for each t, the de Rham cohomology class of the closed differential form

$$
\begin{aligned}
(7.4)\quad ch(A, th) &= \operatorname{trace} \epsilon_F \exp \tilde{F}_{A,th} \\
&= \operatorname{trace} \epsilon_F \exp\left(\tilde{F}_A + \frac{(-1)^{3/4} t}{2\pi} d_A h - \frac{t^2}{2\pi} h^2\right)
\end{aligned}
$$

defines chF. Since the real parts of all eigenvalues of h^2 are positive on C, the "size" of this closed differential form gets smaller exponentially at each point as t tends to get larger. (This is the reason why we put $(-1)^{3/4}$ in the coefficient in the definition of $d_{A,h}$.) If t is sufficiently large, the support of this closed differential form can be said to concentrate "almost" outside of C. This is the localization of the Chern character. From this observation, you may imagine that an element of the relative de Rham cohomology $H^\bullet(X, C)$ is defined using this family of closed differential forms.

We explain two ways of realizing this idea. In each of them, we use closed differential forms on $\mathbb{R} \times X$.

Denote by π_X the projection from $\mathbb{R} \times X$ to X. Define the connection $\tilde{A}$ on the vector bundle $\pi_X^* F$ on $\mathbb{R} \times X$ by the pull-back $\pi_X^* A$. Namely, $d_{\tilde{A}} s = d_A s + d\tau \wedge \partial_\tau s$. Here τ is the coordinate of the $\mathbb{R}$ factor. τh is an endomorphism of $\pi_X^* F$ of degree 1. From these, we obtain the closed differential form $ch(\tilde{A}, \tau h)$ on $\mathbb{R} \times X$.

3.3.1. *The first method of defining the relative Chern character.* Decompose $ch(\tilde{A}, \tau h)$ into the part containing $d\tau$ and the part which does not contain $d\tau$ and write $ch(\tilde{A}, \tau h) = d\tau \wedge \varphi + \psi$. Then the restriction of ψ to $\{\tau_0\} \times X$ coincides with $ch(A, \tau_0 h)$. Since $ch(\tilde{A}, \tau h)$ is a closed differential form, $d_X \varphi = \partial_\tau \psi$ holds. Integrating it over τ from τ_0 to τ_1, we obtain

$$d_X \int_{\tau_0}^{\tau_1} \varphi d\tau = ch(A, \tau_1 h) - ch(A, \tau_0 h).$$

Consider the limit of this equality at each point of a neighborhood of C, when τ_1 tends to infinity. The first term on the right hand side converges to 0 by the expression (7.4). Expressing φ concretely, we find that the "size" of it decays exponentially as τ_1 tends to infinity, hence the integral of the first term converges before taking d_X. Thus

$$\eta_{A,h}(\tau_0) = -\int_{\tau_0}^{\infty} \varphi d\tau$$

is defined as a (smooth) differential form on an open neighborhood of C and satisfies

$$ch(A, \tau_0 h) = d\eta_{A,h}(\tau_0)$$

on the open neighborhood of C. Therefore the element $[ch(A, \tau_0 h), \eta_{A,h}(\tau_0)]$ of $H^\bullet(X, C)$ is defined. We easily see that this element does not depend on the choices of τ_0 and A and that we obtain the same element when we deform h continuously keeping Assumption (C, h) (cf. the following exercise). In particular, if $h^2 = \mathrm{id}_F$ on an open neighborhood of C, $\varphi = 0$ on the open neighbourhood of C, hence this element coincides with the relative Chern character, which we defined previously. We summarize our argument in the following

LEMMA 7.23. *Suppose that h satisfies Assumption (C, h). Let $[h]$ be the homotopy class of h with respect to homotopy keeping Assumption (C, h). Then*

$$ch(F, [h]) = [ch(A, \tau_0 h), \eta_{A,h}(\tau_0)]$$

defines an element of $H^\bullet(X, C)$, which is independent of the choices of A,τ_0 and h (which is a representative of the homotopy class).

EXERCISE 2. Let A_t be a smooth family of connections (commuting with ϵ_F) and h_t a smooth family satisfying Assumption (C, h) parametrized by the real number t. Show that there exist a differential form α on X and a differential form β on an open neighborhood of C such that

$$\begin{aligned} ch(A, \tau_0 h_1) - ch(A, \tau_0 h_0) &= d\alpha, \\ \eta_{A,h_1}(\tau_0) - \eta_{A,h_0}(\tau_0) &= \alpha + d\beta. \end{aligned}$$

(The former equality holds on X, the latter holds on the open neighborhood of C.) Hint: Denote by π_X the projection from $\mathbb{R}^2 \times X$ to X and define the connection $\tilde{A}$ on the pull-back $\pi_X F$ by

$$d_{\tilde{A}} s = d_{A_t} s + d\tau \wedge \partial_\tau s + dt \wedge \partial_t s.$$

Here (τ, t) are coordinates of $\mathbb{R}^2$. Regard $\tilde{h} = \tau h_\tau$ as an endomorphism of $\pi_X^* F$ of degree 1 and think of the closed differential form $ch(\tilde{A}, \tilde{h})$ on $\mathbb{R}^2 \times X$.

3.3.2. *The second method of defining the Relative Chern Character.* Pick and fix a diffeomorphism $u \mapsto \tau = \tau(u)$ from $(-\infty, 0)$ to $\mathbb{R}$ such that $\tau(u) = -u^{-1}$ when u is sufficiently close to 0. Using it, we can regard the closed differential form $ch(\tilde{A}, \tau h)$ on $\mathbb{R} \times X$ as a closed differential form $ch(\tilde{A}, \tau(u)h)$ on $(-\infty, 0) \times X$. We assume that the real part of each eigenvalue of h^2 is positive on an open neighborhood U of C. As τ tends to get larger, $ch(\tilde{A}, \tau h)$ (and its derivatives) decay exponentially at each point of U. Thus we can extend $ch(\tilde{A}, \tau(u)h)$ to a closed differential form, which is continuous up to the boundary, so that it is smooth on the interior of

$$\tilde{X} = (-\infty, 0) \times (X \setminus C) \cup [0, \infty) \times U,$$

which is a subset of $\mathbb{R}_u \times X$. Here, we extend it by 0 on $[0, \infty) \times U$. (This is the same situation as that the function, which is $\exp(-1/x)$ for negative x and 0 for non-negative x, is a smooth function of x.) Denote this extension by the same symbol $ch(\tilde{A}, \tau(u)h)$.

Pick any function ρ on X, which is negative outside of U and positive on C. Denote by P the graph of ρ and regard P as a map from X to $\tilde{X}$. Then we can express the relative Chern character as follows.

LEMMA 7.24. *The pull-back $P^* ch(\tilde{A}, \tau(u)h)$ is a closed differential form, the support of which is disjoint from C, its de Rham cohomology class does not depend on ρ and coincides with $ch(F, [h])$.*

PROOF. We can reduce it to Lemma 7.23. Details are left to readers. □

3.4. Multiplicativity of Relative Chern Character. We explain the multiplicativity of the relative Chern character.

Suppose that h, h' are endomorphisms of the $\mathbb{Z}_2$-graded vector bundles F, F' of degree 1, respectively. Let C, C' be closed subsets of X. We assume that the real part of each eigenvalue of h^2 is positive on each point of an open neighborhood U of C and, similarly, that the real part of each eigenvalue of h'^2 is positive on each point of an open neighborhood U' of C'. By changing h, h' by homotopies, we can assume that the real part of each eigenvalue of h^2, h'^2 is non-negative everywhere on X. From now on, we set this assumption. (For example, if h, h' are Hermitian with respect to some Hermitian inner product, the eigenvalues of h^2, h'^2 are non-negative real numbers.)

Define the endomorphism $\hat{h}$ of $F \otimes F'$ of degree 1 by

$$\hat{h} = h \otimes \mathrm{id}_{F'} + \epsilon_F \otimes h'.$$

Then, at each point of X, each eigenvalue of its square

$$\hat{h}^2 = h^2 \otimes \mathrm{id}_{F'} + \mathrm{id}_F \otimes h'^2$$

is written as the sum of an eigenvalue of h^2 and an eigenvalue of h'^2 at the point. Hence, by the assumption, the real part of each eigenvalue of $\hat{h}^2$ is positive on an open neighborhood of $C \cup C'$ and the relative Chern character $ch(F \otimes F', [\hat{h}])$ is defined as an element of $H^\bullet(X, C \cup C')$.

PROPOSITION 7.25.

$$ch(F \otimes F', [\hat{h}]) = ch(F, [h]) ch(F', [h']).$$

Here, the product on the right hand side is the product

$$H^\bullet(X, C) \times H^\bullet(X, C') \to H^\bullet(X, C \cup C')$$

induced from the exterior product.

PROOF. We give only an outline.

Let π_X be the projection from $\mathbb{R}^2 \times X$ to X. Define the connection $\tilde{A}$ on the vector bundle $\pi_X^*(F \otimes F')$ on $\mathbb{R}^2 \times X$ by the pull-back $\pi_X^*(A \otimes A')$. Namely,

$$\begin{aligned} d_{\tilde{A}}(s \otimes s') = (d_A s) \otimes s' + (-1)^{\deg s} s(d_{A'} s') \\ + d\tau \wedge \partial_\tau (s \otimes s') + d\tau' \wedge \partial_{\tau'}(s \otimes s'). \end{aligned}$$

Here (τ, τ') are the coordinates of $\mathbb{R}^2$. Define the endomorphism $\tilde{h}$ of $\pi_X^*(F \otimes F')$ of degree 1 by $\tilde{h} = \tau h \otimes \mathrm{id}_{F'} + \tau' \epsilon_F \otimes h'$. From these, we define the closed differential form $ch(\tilde{A}, \tilde{h})$ on $\mathbb{R}^2 \times X$.

Let U, resp. U', be the set of all points where the real parts of all eigenvalues of h^2, resp. ${h'}^2$, are positive. They contain C, C', respectively.

As in the previous subsection, we introduce the change of coordinates $\tau = \tau(u)$, $\tau' = \tau'(u')$ and identify $\mathbb{R}^2$ and $(-\infty, 0) \times (-\infty, 0)$ to regard the closed differential form $ch(\tilde{A}, \tilde{h})$ on $\mathbb{R}^2 \times X$ as a closed differential form on $(-\infty, 0) \times (-\infty, 0) \times X$.

Define the subset $\tilde{X}$ of $\mathbb{R}_u \times \mathbb{R}_{u'} \times X$ by

$$\begin{aligned} \tilde{X} &= (-\infty, 0) \times (-\infty, 0) \times (X \setminus (C \cup C')) \\ &\quad \cup [0, \infty) \times \mathbb{R} \times U \cup \mathbb{R} \times [0, \infty) \times U'. \end{aligned}$$

Using the fact that the real part of each eigenvalue of h^2, ${h'}^2$ are non-negative everywhere, we extend the closed differential form $ch(\tilde{A}, \tilde{h})$ on $(-\infty, 0) \times (-\infty, 0) \times X$ by 0 to regard it as a closed differential form, which is smooth on the interior of $\tilde{X}$ and continuous up to (the portion of) the boundary of $\tilde{X}$ (contained in $\tilde{X}$).

When P is a section of the fiber bundle $\mathbb{R}^2 \times X \to X$ and its image is contained in $\tilde{X}$, the pull-back of $ch(\tilde{A}, \tilde{h})$ by P is a closed differential form on X with support being disjoint from $C \cup C'$. Such P's are homotopic to each other. Hence the elements of $H^\bullet(X, C \cup C')$ defined by their pull-backs are the same.

Pick the functions ρ, which is negative outside of U and positive on C, and ρ', which is negative outside of U' and positive on C'. Also pick another function $\hat{\rho}$, which is negative outside of $U \cup U'$ and positive on $C \cup C'$. Using them, define sections P_0, P_1, the images of which are contained in $\tilde{X}$, by

$$P_0(x) = (x, \hat{\rho}(x), \hat{\rho}(x)), \quad P_1(x) = (x, \rho(x), \rho'(x)).$$

Then, using Lemma 7.24, we find that

$$[P_0^* ch(\tilde{A}, \tilde{h})] = ch(F \otimes F', [\hat{h}]), \quad [P_1^* ch(\tilde{A}, \tilde{h})] = ch(F, [h]) ch(F', [h']).$$

The desired multiplicativity follows from this.

□

We can easily prove the following lemma using the multiplicativity of the relative Chern character.

LEMMA 7.26. *For an oriented Euclidean space E of even dimension, we can construct the pair $(E \times \Delta_E, [\sqrt{-1}\, c_E])$ on E from the spinor representation space Δ_E and its Clifford multiplication c_E associated with E. This gives the element $ch(E \times \Delta_E, [\sqrt{-1}\, c_E])$ of $H_c^\bullet(E)$. Then its integral*

$$\int_E ch(E \times \Delta_E, \sqrt{-1}\, c_E)$$

over E is equal to 1.

PROOF. We give an outline. When $E = E_0 \oplus E_1$ with oriented even dimensional Euclidean spaces E_0, E_1, $(E \times \Delta_E, \sqrt{-1}\, c_E)$ is the product of $(E_0 \times \Delta_{E_0}, \sqrt{-1}\, c_{E_0})$ and $(E_1 \times \Delta_{E_1}, \sqrt{-1}\, c_{E_1})$, hence, by the multiplicativity of ch, we obtain

$$\begin{aligned}
&\int_E ch(E \times \Delta_E, \sqrt{-1}\, c_E) \\
&\quad = \int_{E_0} ch(E_0 \times \Delta_{E_0}, \sqrt{-1}\, c_{E_0}) \int_{E_1} ch(E_1 \times \Delta_{E_1}, \sqrt{-1}\, c_{E_1}).
\end{aligned}$$

Thus the problem is reduced to the case that E is 2 dimensional. In this case, we can see the claim similar to the proof of Lemma 7.9 (cf. Question 1). □

4. Thom Class and Thom Isomorphism

Firstly, we review the Thom isomorphism for de Rham cohomology. For the purpose of this section, we do not need the whole theory of the Thom isomorphism; the notion of "integration along fibers", which we explained in Chapter 1, is sufficient. However, when we see that the "Thom isomorphism in K-theory" is closely related to the index theorem in Chapter 8 of the sequel, it makes understanding easier if we compare it with the prototypical argument. Thus we will give it in this section.

Let $\pi_F : F \to X$ be a fiber bundle on X with orientable fibers. To exclude complication of notation, we assume that X is a closed manifold.

LEMMA 7.27. *The integration along fibers commutes with the exterior derivative up to sign.*

We briefly explain the reason why this lemma holds in the following remark and omit its proof.

REMARK 7.28. A differential form of degree k is characterized by assigning the number defined by the integration to each (very small) k-dimensional subspace of X. For a differential form α on F, its integration along fibers corresponds to the map which sends a (very small) subspace Δ of X to the integration of α over the inverse image $\pi_F^{-1}(\Delta)$. Namely, the integration along fibers is the "dual of the map taking the inverse image". The exterior derivative is considered as the dual of the map, which sends a (very small) r-dimensional subspace to its boundary, a $(r-1)$-dimensional subspace (namely, Stokes' theorem). The inverse image $\pi_F^{-1}(\partial\Delta)$ of the boundary of Δ and $\partial\pi_F^{-1}(\Delta)$, the boundary of the inverse image, coincide as geometric objects. This is the essential reason why the integration along fibers commutes with the exterior derivative up to sign.

Thus, using the integration along fibers, we can define the map

$$\int_{\pi_F : F \to X} : H_c^\bullet(F) \to H^\bullet(X, \theta_{\pi_F}),$$

which becomes a homomorphism as $H^\bullet(X)$-modules (cf. Chapter 2, §1). Here $H_c^\bullet(F)$ is the de Rham cohomology obtained from the exterior derivative on the space of all differential forms with compact support.

For any local system θ on X, we can extend the integration along fibers to

$$\int_{F \to X} : \Omega_c^\bullet(F, \theta_{\pi_F} \otimes \pi^*\theta) \to \Omega^\bullet(X, \theta).$$

Similarly, we obtain the map between corresponding de Rham cohomologies. We prove the following theorem at the end of this section.

THEOREM 7.29 (Thom Isomorphism Theorem). *Let F be a real vector bundle on X of rank r and θ any local system on X. Denote* *by θ_{π_F} the local system of orientations along fibers of F. The $H^\bullet(X)$-module homomorphism*

$$\int_{F\to X} : H_c^\bullet(F, \theta_{\pi_F} \otimes \pi_F^* \theta) \to H^\bullet(X, \theta)$$

induced by the integration along fibers is an isomorphism.

The inverse map of the integration along fibers is given by the multiplication with the Thom class, which is defined as follows.

DEFINITION 7.30. Let F be a real vector bundle on a closed manifold X of rank r. If the value

$$\int_{F\to X} \tau \in H^0(X) \cong \mathbb{R}$$

of the integration along fibers of an element τ of $H_c^r(F, \theta_{\pi_F})$ is 1, we call τ the **Thom class** of F.

LEMMA 7.31. *The Thom class exists.*

PROOF. By Theorem 2.5 in Chapter 2 (adapted for real vector bundles), there exists a projection $p : X \times \mathbb{R}^N \to F$ from a trivial vector bundle to F. The Thom class τ_0 for the trivial vector bundle can be immediately constructed from an element of $\Omega_c^N(\mathbb{R}^N)$, the integral of which is 1. Then the integration along fibers of p

$$\int_{p:X\times\mathbb{R}^N\to F} \tau_0$$

gives the desired Thom class of F. □

ANOTHER PROOF. Firstly, we consider the case that the rank r of F is even and that F is equipped with an orientation and carries a $Spin^c$-structure. Denote by S_F the $\mathbb{Z}_2$-graded spinor bundle associated with this $Spin^c$-structure. Then we find that

$$ch_{r/2}(\pi_F^* S_F, \sqrt{-1}\, c_{S_F})$$

enjoys the desired property and is one of the Thom classes by Lemma 7.26.

Next, we consider the general case. The rank of $F \oplus F \cong F \otimes \mathbb{C}$ is the even number $2r$ and equipped with the orientation. It also carries a $Spin^c$-structure. Denote by W_F the corresponding spinor bundle to find that $W_F \cong \bigwedge^\bullet F \otimes \mathbb{C}$. Then $ch(\pi_{F\oplus F}^* W_F, \sqrt{-1}\, c_{W_F})$ is a Thom class of $F \oplus F$. To obtain a Thom class of F from this class, it is enough to consider, for example, the projection $p_1 : F \oplus F \to F$ to the first factor and the integration along the fibers of p_1

$$\int_{p_1 : F\oplus F\to F} ch(\pi^*_{F\oplus F} W_F, \sqrt{-1}\, c_{W_F}).$$

$\square$

Concrete construction of W_F and c_{W_F} in the preceding proof goes as follows.

Denote by $\tilde{\pi}$ the projection from $F \oplus F$ to X. The Clifford multiplication on the $\mathbb{Z}_2$-graded vector bundle $W_{F\oplus F} = \tilde{\pi}^* \bigwedge^\bullet F \otimes \mathbb{C}$ on $F \oplus F$ is given by

$$c_{W_F}(u, v) = (u^\wedge - u^\lrcorner) + \sqrt{-1}\,(v^\wedge + v^\lrcorner)$$

for an element (u, v) of $F \oplus F$. (Here we identified F and F^* using the Euclidean metric on F.)

This is closely related to the family of supersymmetric harmonic oscillators defined in Chapter 6, §3. Denote by

$$(W_{F,\mathrm{fiber}}, D_{F,\mathrm{fiber}}, h_{F,\mathrm{fiber}})$$

the family of supersymmetric harmonic oscillators along fibers of F. Then

$$\pi_F^* W_F \cong W_{F,\mathrm{fiber}}$$

and we can identify by a suitable isomorphism

$$\sqrt{-1}\, c_{W_F} \cong \sqrt{-1}\, c_{W_{F,\mathrm{fiber}}} + t h_{F,\mathrm{fiber}}$$

for $t > 0$.

Using existence of the Thom class, we prove the Thom isomorphism theorem.

PROOF OF THEOREM 7.29. It is enough to show

$$a = \left(\int_{F\to X} a\right) \tau_F$$

for any element a of $H_c^\bullet(F, \theta)$.

Consider the vector bundle $F \oplus F$ on X. Denote by $p_1, p_2 : F \oplus F \to F$ the map taking the fiberwise projections to the first factor and the second factor, respectively.

By the operation of taking the product

$$H_c^\bullet(F, \pi_F^*\theta) \times H_c^\bullet(F, \theta_{\pi_F}) \to H_c^\bullet(F \times F, p_1^*\pi_F^*\theta \otimes p_2^*\theta_{\pi_F}),$$

we take the product $a \cdot \tau_F$ and consider its restriction to the closed subset $F \oplus F$. Taking the integration along fibers of p_1, resp. p_2, we obtain the right hand side, resp. the left hand side, of the above equality. On the other hand, p_1 and p_2 are joined by the homotopy $p_1 \cos s + p_2 \sin s$ $(0 \leq s \leq \pi/2)$. Thus we find the coincidence of the right hand side and the left hand side. □

REMARK 7.32.

(1) By the Thom isomorphism theorem, we especially find the uniqueness of the Thom class. Denote the Thom class of F by $\tau(F)$.
(2) The proof of the Bott periodicity goes in a way parallel to the above argument (use K-groups instead of de Rham cohomologies, and use the index for families instead of the integration along fibers).

5. The Euler Class

In this book, we define the Euler class using the Thom class.

DEFINITION 7.33. We call the image of $\tau(F)$ by the degree preserving map

$$H_c^\bullet(F, \theta_{\pi_F}) \to H^\bullet(X, \theta_{\pi_F}|X),$$

which is induced from the restriction to the zero section, the **Euler class** and denote it by $e(F)$.

We give a concrete description of the Euler class.

Let $\pi : F \to X$ be a real vector bundle equipped with a Euclidean metric on a closed manifold X of rank r.

Denote by (W_F, h_F) the family of pairs associated with the supersymmetric harmonic oscillators along fibers of F. W_F is a vector bundle on F and carries a structure of the $\mathbb{Z}_2$-graded Clifford module over $T_{\text{fiber}}F$. Pick an isomorphism $T_{\text{fiber}}F \cong F \oplus F$ and denote the diagonal map by

$$\text{diag}_F : \ F \to F \oplus F \cong T_{\text{fiber}}F.$$

Then the degree r components of the two elements

$$\int_\pi ch(W_F, h_F), \quad \int_\pi ch(W_F, \sqrt{-1}\, c_F(\text{diag}_F))$$

is an element of $H^r(X, \theta_{\pi_F}|X)$.

PROPOSITION 7.34. *Each of the degree r components above coincides with the Euler class $e(F)$. The components of degree less than r are 0.*

PROOF. $F \oplus F \cong F \otimes \mathbb{C}$ carries a $Spin^c$-structure. Denote by W_F the spinor bundle associated with it. Then the Thom class of F is given by the degree r components of

$$\int_{p_i : F \oplus F \to F} ch(\pi^*_{F \oplus F} W_F, \sqrt{-1}\, c_{W_F}(\mathrm{diag}_{F \oplus F}))$$

for the projections $p_1, p_2 : F \oplus F \to F$ to the first and second factor, respectively. The degree k component is an element of $H^k_c(F \oplus F)$, which is 0 for $k < r$ by the Thom isomorphism. Restricting the degree r component to the zero section X, we obtain the Euler class by the definition. □

If F is of rank $2m$, equipped with the orientation, and, in addition, carries a $Spin^c$-structure, the Thom class is easily constructed without using the integration along fibers. Hence the Euler class is also easily described.

PROPOSITION 7.35.

(1) *Suppose that F is of rank $2m$, equipped with the orientation and, in addition, carries a $Spin^c$-structure, Denote by S the corresponding spinor bundle to obtain $e(F) = ch_m(S)$.*
(2) *For complex line bundles $L_1, L_2, \ldots, L_m$ on X, write $x_k = ch_1(L_k)$. Then we have $e(L_1 \oplus L_2 \oplus \cdots \oplus L_m) = x_1 x_2 \cdots x_m$.*
(3) *More generally, when F is a complex vector bundle of rank m, we have $e(F) = c_m(F)$.*

PROOF. The first claim clearly follows from the construction in Lemma 7.31 and the definition of the Euler class.

Using it, we prove the second claim. For the $Spin^c$-structure determined by the complex vector bundle structure of L_k, the corresponding spinor bundle S_k is

$$S_k = \bigwedge^0 \bar{L}_k \oplus \bigwedge^1 \bar{L}_k = \mathbb{C} \oplus \bar{L}_k$$

as a $\mathbb{Z}_2$-graded vector bundle. Then we have

$$
\begin{aligned}
(7.5) \qquad ch(S_1 \otimes \cdots \otimes S_m) &= \prod_{k=1}^{m} ch(S_k) \\
&= \prod_{k=1}^{m} (ch(\mathbb{C}) - ch(\bar{L_k})) \\
&= \prod_{k=1}^{m} (1 - e^{-x_k}) \\
&= \prod_{k=1}^{m} (x_k + O(x_k^2))\,.
\end{aligned}
$$

Hence its degree $2m$ component is $\prod_{k=1}^{m} x_k$.

The last claim can be reduced to $ch_m(\bigwedge^\bullet \bar{F}) = c_m(F)$.

A connection A on F naturally induces the connection A_S on $S = \bigwedge^\bullet \bar{F}$. Here the curvature F_{A_S} of A_S is determined only by the curvature F_A of A. Hence we can describe $ch_m(A_S)$ in terms of F_A. The concrete description is as follows.

For a commutative algebra R over $\mathbb{C}$, we can define the map

$$
\begin{aligned}
\Phi : R \otimes \mathrm{End}(\mathbb{C}^m) &\to R \\
a &\to \mathrm{trace}\, \epsilon_{\mathrm{End}(\bigwedge^\bullet \overline{\mathbb{C}^m})} \frac{1}{m!} (\phi a)^m
\end{aligned}
$$

using the natural map $\phi : \mathrm{End}(\mathbb{C}^m) \to \mathrm{End}(\bigwedge^\bullet \overline{\mathbb{C}^m})$. Here Φ is a polynomial of m^2 elements of R with rational coefficients.

At each point x of X, fix any trivialization $(F)_x \cong \mathbb{C}^m$ of F and consider $R = \bigoplus_k \bigwedge^{2k} (T^*X)_x$. Then we obtain $ch(A_S)_x = \Phi(\tilde{F}_A)_x$. On the other hand, we have, by the definition, $c_m(A)_x = \det(\tilde{F}_A)_x$. Hence it is sufficient for showing the claim to check the coincidence of the polynomial Φ with rational coefficients and det. This is purely a proposition in algebra.

In order to show the coincidence as polynomials with rational coefficients, it is enough to show it in the case of $R = \mathbb{C}$. By the definition of Φ, the images of two elements which are conjugate to each other by Φ are the same. Hence it is enough to check $\Phi(a) = \det a$ for diagonal matrices a. It is easy to compute this directly. $\square$

In fact, the methods of the proofs of the last claim (3) and the second claim (2) concerning the direct sum of line bundles are related.

The computation showing $\Phi(a) = \det a$ for diagonal matrices is parallel to the computation showing (2) (this is the reason why we omit the computation).

In many cases, it is enough for proving equalities among characteristic classes of complex vector bundles (and its associate bundles) to establish the equalities for the case that the vector bundles are split into the direct sums of line bundles through the argument as in the proof of (3). Such an argument is called the **splitting principle**.

REMARK 7.36. It is possible to formulate the splitting principle purely in the realm of topology. Usually, we rather call the topological counterpart the splitting principle.

SUMMARY.

7.1 We defined the Chern character and Chern classes for complex vector bundles as elements of the de Rham cohomology using the curvature of the connection.

7.2 For $\mathbb{Z}_2$-graded vector bundles with compact support, we defined the relative Chern character as elements of the relative de Rham cohomology using a variant of "localization".

7.3 For the de Rham cohomology of the vector bundles (consisting of differential forms of compact support along fibers), the operation of taking the integral along fibers and the operation of taking the multiplication with the Thom class are inverse to each other.

7.4 We defined the Euler class as the restriction of the Thom class to the zero section.

Index